Best Practices in Developing Proactive Supply Strategies for Air Force Low-Demand Service Parts

Mary E. Chenoweth, Jeremy Arkes, Nancy Y. Moore

Prepared for the United States Air Force

PROJECT AIR FORCE

The research described in this report was sponsored by the United States Air Force under Contracts FA7014-06-C-0001 and F49642-01-C-0003. Further information may be obtained from the Strategic Planning Division, Directorate of Plans, Hq USAF.

Library of Congress Cataloging-in-Publication Data

Chenoweth, Mary E.
  Best practices in developing proactive supply strategies for Air Force low-demand service parts / Mary E. Chenoweth, Jeremy Arkes, Nancy Y. Moore.
    p. cm.
  Includes bibliographical references.
  ISBN 978-0-8330-4878-3 (pbk. : alk. paper)
  1. United States. Air Force–Aviation supplies and stores. 2. United States. Air Force–Procurement. 3. United States. Air Force–Equipment–Maintenance and repair. 4. Airplanes, Military–United States–Maintenance and repair I. Arkes, Jeremy. II. Moore, Nancy Y., 1947- III. Title.

  UG1103.C48 2010
  358.4'18–dc22

                                                                                        2010016541

The RAND Corporation is a nonprofit research organization providing objective analysis and effective solutions that address the challenges facing the public and private sectors around the world. RAND's publications do not necessarily reflect the opinions of its research clients and sponsors. **RAND®** is a registered trademark.

Cover image courtesy of the U.S. Air Force.

© Copyright 2010 RAND Corporation

Published 2010 by the RAND Corporation
1776 Main Street, P.O. Box 2138, Santa Monica, CA 90407-2138
1200 South Hayes Street, Arlington, VA 22202-5050
4570 Fifth Avenue, Suite 600, Pittsburgh, PA 15213-2665
RAND URL: http://www.rand.org/
To order RAND documents or to obtain additional information, contact
Distribution Services: Telephone: (310) 451-7002;
Fax: (310) 451-6915; Email: order@rand.org

Preface

The Air Force uses thousands of different items to support its aircraft. Most of these seldom need replacement. In fact, of the nearly 60,000 different items identified in an analysis of annual requisitions of weapon system components, nearly three in four had no more than a half-dozen annual requisitions. But just as difficulties obtaining more frequently ordered parts can affect aircraft availability, so can difficulties obtaining low-demand parts. Furthermore, managing such parts can be more difficult and expensive on average. For example, their unit costs are often higher because of such fixed costs as physical plants, manufacturing and repair equipment, and because overhead is apportioned over a smaller total quantity of parts.

This monograph reviews Air Force purchases of "low-demand" parts, analyzing how much the Air Force spends on low-demand parts and the types of parts that have a low demand. It then identifies and synthesizes best commercial purchasing and supply chain management practices used for developing supply strategies for such items. Finally, the monograph recommends how the Air Force could improve its supply strategies for such items.

This research was part of a study on "Best Practices for Purchasing and Supply Chain Management: Developing Effective Market Research Methods and Proactive Supply Strategies for Low-Demand Items," sponsored by the Director of Transformation, Deputy Chief of Staff for Logistics, Installations, and Mission Support, Headquarters U.S. Air Force, Washington, D.C. (AF/A4I), and the Deputy Assistant Secretary for Contracting, Office of the Assistant Secretary of the Air

Force for Acquisition (SAF/AQC). It was conducted in the Resource Management Program of RAND Project AIR FORCE.

This monograph should be of interest to anyone interested in purchasing and supply chain management for the Department of Defense, particularly for the air logistics centers and the Defense Logistics Agency. This research may assist Commodity Councils of the Air Force Materiel Command in exercising their commandwide responsibilities for developing supply strategies for selected groups of commodities and weapon system acquisition program offices and life-cycle managers.

Related work for the Air Force has been documented in

- Jeremy Arkes and Mary E. Chenoweth, *Estimating the Benefits of the Air Force Purchasing and Supply Chain Management Initiative,* Santa Monica, Calif.: RAND Corporation, MG-584-AF, 2007
- Nancy Y. Moore, Clifford A. Grammich, and Robert Bickel, *Developing Tailored Supply Strategies,* Santa Monica, Calif.: RAND Corporation, MG-572-AF, 2007
- Nancy Nicosia and Nancy Y. Moore, *Implementing Purchasing and Supply Chain Management: Best Practices in Market Research,* Santa Monica, Calif.: RAND Corporation, MG-473-AF, 2006
- Mary E. Chenoweth and Clifford A. Grammich, *F100 Engine Purchasing and Supply Chain Management Demonstration: Findings from Air Force Spend Analyses,* Santa Monica, Calif.: RAND Corporation, MG-424-AF, 2006
- Lloyd Dixon, Chad Shirley, Laura H. Baldwin, John A. Ausink, and Nancy F. Campbell, *An Assessment of Air Force Data on Contract Expenditures,* Santa Monica, Calif.: RAND Corporation, MG-274-AF, 2005
- Nancy Y. Moore, Cynthia R. Cook, Clifford A. Grammich, and Charles Lindenblatt, *Using a Spend Analysis to Help Identify Prospective Air Force Purchasing and Supply Management Initiatives: Summary of Selected Findings,* Santa Monica, Calif.: RAND Corporation, DB-434-AF, 2004
- Nancy Y. Moore, Laura H. Baldwin, Frank Camm, and Cynthia R. Cook, *Implementing Best Purchasing and Supply Management*

Practices: Lessons from Innovative Commercial Firms, Santa Monica, Calif.: RAND Corporation, DB-334-AF, 2002
- Laura H. Baldwin, Frank Camm, and Nancy Y. Moore, *Federal Contract Bundling: A Framework for Making and Justifying Decisions for Purchased Services,* Santa Monica, Calif.: RAND Corporation, MR-1224-AF, 2001
- Laura H. Baldwin, Frank Camm, and Nancy Y. Moore, *Strategic Sourcing: Measuring and Managing Performance,* Santa Monica, Calif.: RAND Corporation, DB-287-AF, 2000.

RAND Project AIR FORCE

RAND Project AIR FORCE (PAF), a division of the RAND Corporation, is the U.S. Air Force's federally funded research and development center for studies and analyses. PAF provides the Air Force with independent analyses of policy alternatives affecting the development, employment, combat readiness, and support of current and future aerospace forces. Research is conducted in four programs: Force Modernization and Employment; Manpower, Personnel, and Training; Resource Management; and Strategy and Doctrine.

Additional information about PAF is available on our website: http://www.rand.org/paf/

Contents

Figures

Tables

Summary

The Air Force uses thousands of parts to maintain its aircraft. While some of these are replaced quite often, most of them, in fact, are needed infrequently. Of nearly 60,000 items the Air Force requisitioned in a recent three-year period, nearly three in four had no more than a half-dozen annual requisitions.

Difficulties in developing effective supply strategies for these "low-demand" parts can cause problems in maintaining aircraft just as problems in purchasing and supply chain management for more frequently used parts can. In fact, the unpredictability of needs for low-demand parts, as well as the difficulties in attaining them from suppliers who find them less lucrative to produce than more commonly used parts, can mean that they pose even more challenges for Air Force operations than do parts with higher demands. Low-demand parts also pose challenges to Air Force purchasing and supply chain management goals, such as those for reducing sourcing cycle time, improving supply material availability, and decreasing material purchase and repair costs, particularly when low-demand parts are used by larger, more expensive parts and assemblies.

To assess how the Air Force could improve its development of supply strategies for low-demand parts, we performed three tasks. First, we reviewed business, academic, and defense literature on best practices for developing supply strategies for low-demand parts. Second, we analyzed spend data on these parts, including some indicators of which low-demand parts may affect mission capability for aircraft or incidents of awaiting parts. Third, we interviewed representatives of leading commercial firms regarding their best practices for developing supply strategies for these kinds of parts. Below we discuss data on low-demand parts used by the Air Force, best commercial practices for obtaining these parts from private-sector suppliers, and their applicability to the Air Force.

Low-Demand Parts in the Air Force

We defined low-demand parts as those that had no more than six average annual requisitions. In our interviews with commercial companies, we found that frequency of requisition was both the most common criterion for determining whether a part was "low-demand" and the variable of greatest concern to managing "low-demand" issues.

We found that Air Force low-demand parts are concentrated in relatively few classes of goods. Nearly two-thirds are in just five federal supply groups, and nearly one-third are in the federal supply group for electric and electronic equipment components. Air logistics centers rather than Air Force bases are the most frequent users of low-demand items. Nearly one-half of the requisitions for low-demand items originated at the Warner Robins, Oklahoma City, or Ogden Air Logistics Centers. This suggests that much improvement may result from initiatives focused on a select group of items at a small number of locations.

Spend data on low-demand items also indicate some additional commodities warranting attention. Most of the dollars the Air Force spends on low-demand items are for items that are in the same classes as items that consume most of the Air Force's dollars. For example, engines, turbines, and components form the federal supply group in which the Air Force spends both the most money overall for spare and repair items and the most money for low-demand items. Many leading suppliers of low-demand items are also leading suppliers of all spare and repair items to the Air Force. Of the top ten suppliers to the Air Force, eight are also among the top ten suppliers of low-demand items. This indicates that many of the suppliers with which the Air Force should implement overall supply strategies are the same as those with which it should address supply strategies for low-demand items. The relatively greater proportion of low-demand items that are currently purchased through purchase orders rather than contracts compared to non-low-demand items indicates that such items are less likely to have supply strategies already in place for them.

Requisition data was used to define low-demand parts as anything that had six or fewer demand occurrences per year. The match rate between the requisition and "mission capable" (MICAP) and awaiting parts (AWP) data was very low, which meant low-demand status of MICAP and AWP parts could not be based on requisition data. To assess the extent to which such supply availability problems for MICAP and AWP parts were attributable to low-demand parts, we considered as low-demand parts insurance (INS) items with no projected demands and nonstockage objective (NSO) items with very low projected demands. We found that of all items associated with MICAP or AWP incidents, one in four were INS or NSO items, and of all MICAP or AWP incidents, about one in fourteen incidents involved an INS or NSO item.

Best Practices for Developing Supply Strategies for Low-Demand Items

Where a product is in its life cycle determines to a large degree the best practices that are available to assure the supply of low-demand parts for it. Our interviewees outlined strategies for ensuring supply during the three life-cycle phases of design, production, and postproduction. The best options for ensuring long-term aftermarket support are best put in place in the earliest stages of a product's life cycle.

The first, design, phase of a product's life cycle can offer the most opportunities for minimizing the total number of low-demand parts. Involving buyers and suppliers in the design of new systems, products, and parts can help balance performance and cost objectives while minimizing unique parts to a specific product. Reducing complexity by using common subsystems and parts can also help minimize unique parts and hence the ultimate number of parts that may have only low demands. Monitoring and managing obsolescence through a product's life cycle can help avoid the need to replace parts that manufacturers or suppliers might discontinue.

The second, production, phase is the best time to align supplier incentives with goals for long-term support. In fact, buyers' leverage over suppliers peaks just before award of the production contract. Committing suppliers to postproduction aftermarket services in the production contract can help ensure long-term support of low-demand parts throughout the life cycle of a product. Buyers may also want to leverage the production contract award to ensure potential access to technical data for low-demand parts, which can help it develop alternative sources of supply should the original supplier exit the business or prove to be unresponsive.

The third, postproduction, phase offers fewer good options for developing supply strategies. Buyers seeking to ensure supply of low-demand parts may wish to provide incentives for supply of low-demand parts so as to maintain continuity in the supply chain. If a supplier has left the business, buyers may choose to develop a new one. Buyers may also choose to buy a lifetime supply of parts before a supplier exits a business, although it can be difficult to estimate the quantities needed for a "lifetime" supply. Finally, buyers may purchase or retire whole products to cannibalize low-demand (and other) parts on them.

Options for the Air Force

Just as best practices for supply strategies regarding low-demand parts depend on where a product is in its life cycle, so do the options available to the Air Force. In assessing supply strategies for low-demand parts, the Air Force must also consider many conditions that differ from those in the private sector, including the presence of fewer suppliers and hence of alternative sources of supply. We offer the following recommendations.

In the design phase of a product, the Air Force should encourage active participation of sustainment personnel who are able to incorporate reliability and maintainability concerns (see pp. 37–38, 44–45, 58, 60–61, 65, 72). These personnel could work more effectively with acquisition personnel to plan for long-term parts support and help ensure greater consideration for supply chain issues throughout a product's life cycle.

Having the production contract include language that requires the supplier to provide long-term support can help align supplier incentives with Air Force goals and make commitment to them more likely to be honored (see pp. 28–29, 44–45, 58, 64–65, 71–72). The Air Force should work with suppliers to put in place strategies for low-demand parts support during production, when problems are easier to address than they are in postproduction, when lower-tier suppliers may quit the business (see pp. 35–36, 38–39, 45–46, 58, 62, 66–67, 72–73). During production, the Air Force should also seek access or an agreement to possibly access technical data for parts that would help it find alternative sources of supply if necessary after production ends (see pp. 43–44, 48–50, 58, 72).[1]

Air Force options for support and supply strategies following production are limited. The Air Force could analyze its low-demand parts to determine groups of parts that ought to be purchased together to make support of them more attractive to suppliers (see pp. 53–55, 58, 66–69, 73). In some cases, the Air Force may benefit from working with other buyers, such as the Navy or Army, in exercising leverage over suppliers to improve support for low-demand parts (see pp. 52, 67–68, 74).

The Air Force operates many legacy aircraft, which means that most of its fleet is postproduction and thus facing technology obsolescence, diminishing sources of supply and repair, and more low-demand failures than younger fleets. Few systems are still in the preaward phase (e.g., the replacement tanker). However, many subsystems continue to be modernized and upgraded. To the extent that these programs are competed, opportunities exist to apply some of these principles even to legacy aircraft (see pp. 39–42, 58, 61–62, 71, 73–74).

---

[1]   Total Life Cycle Cost Systems Management, which requires that sustainment costs be considered more explicitly in the production phase of the acquisition of a weapon system, could make the buying of technical data more likely than in the recent past. Programs often reallocate monies earmarked for technical data to cover cost overruns during development and/or production, which essentially derails plans to buy technical data.

Acknowledgments

We thank the companies and individuals who participated in this study. Several companies expended considerable time and resources in making senior personnel available for discussions. Because we ensured the anonymity of participants, we cannot identify them by name, but we appreciate all of those who contributed to this study by sharing with us their broad experiences with supply strategies of low-demand parts.

We wish to acknowledge others who also participated in this study. Personnel at Warner Robins Air Logistics Center met with us to discuss the practices the Air Force uses for developing supply strategies for low-demand parts. Air Force Security Assistance Center personnel met with us to describe the Parts and Repair Ordering System, which provides spare parts support of Foreign Military Sales aircraft, and how it works.

RAND colleague and project member Nancy Nicosia set up several interviews and participated in them. Shelley Wiseman helped us organize this material, and Clifford Grammich streamlined and clarified the text. Robert Tripp reviewed an early draft, which led to further improvements.

Finally, we thank Jim Masters and Mark Wang for their thorough and thoughtful reviews of the draft report.

Abbreviations

AFGLSC	Air Force Global Logistics Command
AFMC	Air Force Materiel Command
ALC	Air Logistics Center
AWP	awaiting parts
CLS	Contractor Logistics Support
COTS	commercial-off-the-shelf
D&SWS	Develop and Sustain Warfighting Systems
DLA	Defense Logistics Agency
DMEA	Defense Microelectronics Activity
DoD	Department of Defense
DoDAAC	Department of Defense Activity Address Code
FAR	Federal Acquisition Regulation
FSC	Federal Supply Class
FSG	Federal Supply Group
FY	fiscal year
GCSS-AF	Global Combat Support System–Air Force
GIDEP	Government-Industry Data Exchange Program
INS	insurance
MICAP	mission capable
NIIN	National Item Identification Number
NSO	nonstockage objective
OEM	original equipment manufacturer

PBL   Performance-Based Logistics
UAV   unmanned aerial vehicle
UMMIPS   Uniform Material Movement and Issue Priority
             System
UTC   United Technologies Corporation

CHAPTER ONE

Introduction

The Air Force uses thousands of service parts to maintain its aircraft. It uses these items to replace those removed from fielded aircraft or equipment when they fail or are scheduled to be removed because of the time or extent of their usage. For parts that are routinely removed and replaced, the Air Force stocks an inventory capable of meeting most demands. Several recent initiatives have sought to improve purchasing and supply chain management of the parts in greatest demand accounting for most dollars spent and having the greatest effect on replacement and repairs.

The Air Force is similar to many private enterprises in that while it uses thousands of different parts, most of them are needed infrequently. Of nearly 60,000 items the Air Force requisitioned between fiscal year (FY) 2002 and FY 2004, nearly three in four had no more than a half-dozen annual requisitions, indicating low demand for them.[1]

Nevertheless, difficulties in managing these "low-demand" parts can affect operations just as difficulties in attaining more frequently ordered parts can.[2] Difficulties in managing the supply of low-demand

---

[1] Most Air Force–managed parts have no demands in a given year. From FY 2006 to FY 2008, only about 37 percent of the National Item Identification Numbers (NIINs) in the D200 database used to forecast requirements had demands. (On average, 43,873 different NIINs were ordered each year compared to the average 120,184 NIINs in the D200 database.) These dormant parts can arise from variable demands with low-demand items. Data sources: Requisition data extracted from the Strategic Sourcing Analysis Tool, Global Combat Support System–Air Force, and D041 USAF Recoverable Spares Data Base, March of each year.

[2] In the private sector, "low-demand" parts can also be called "slow-moving," "non-routine," or "unscheduled."

parts led the Director of Transformation, Deputy Chief of Staff for Logistics, Installations, and Mission Support, Headquarters U.S. Air Force, Washington, D.C. (AF/A4I), and the Deputy Assistant Secretary for Contracting, Office of the Assistant Secretary of the Air Force for Acquisition (SAF/AQC) to ask RAND Project AIR FORCE to identify best practices used by companies in developing supply strategies for purchasing low-demand parts in cost-effective and efficient ways. At the time of this study, Air Force commodity councils were developing supply strategies for the bulk of their demands, i.e., items with high to medium demands. Best practices for these kinds of items were well-established in the private sector. Taking root in the private sector were best practices in developing supply strategies for low-demand items. This interest was fueled in part by a shift toward performance-based services, in which suppliers were paid for keeping equipment running rather than individual repairs and spares.[3] Once suppliers were accountable for delivering equipment performance rather than parts, then any part that caused equipment to stop operating needed to be supported, including low-demand items.

Difficulties in managing low-demand parts can affect both specific aircraft operations as well as more general management of resources. The C-5 cargo aircraft provides one example of the impact that low-demand parts availability can have on readiness. An Air Force Materiel Command (AFMC) analysis found that 76 percent of the items considered as mission capable (MICAP) drivers (either in the top 20 percent of MICAP hours or incidents) were those with fewer than ten demands per year. This same analysis found that low-demand parts made up about 80 percent of all C-5 parts and 90 percent of C-5 reparable parts in demand forecasting models.[4] The difficulties in stock-

---

[3] In 2002 about 40 percent of the $10.6 billion in sales for GE Aircraft Engines was engine parts and services and as much as two-thirds of its $2.1 billion in operating profit (Siekman, 2002). An early example of the paradigm shift from buying repairs and spares from vendors to services is Intel and its Supplier Support Program, which paid suppliers more if their equipment operated past set targets (Morgan, 1995).

[4] Of all parts needed to restore MICAP status for the C-5, 54 percent had no field stock levels. As a result, when parts failed at locations with no stock, MICAP incidents resulted if parts could not be repaired locally and quickly returned to the aircraft. MICAP incidents

ing and managing low-demand parts contributed to mission capable rates of only about 60 percent for this aircraft (AFMC, 2001; Warner Robins Air Logistics Center, 2000).

Low-demand parts are less likely to be stocked in base supply, which means when needed they must first be ordered or requisitioned from a wholesale supply source such as an Air Logistics Center (ALC) or the Defense Logistics Agency (DLA). Low-demand parts can delay repairs of other, larger components at the intermediate or depot level if a particular part needed in the repair process is not stocked or available.

Low-demand parts can also pose higher costs in purchasing and supply chain management. Like all parts, low-demand items carry at least two types of cost: the direct procurement or manufacturing cost and a potential accounting cost for material that does not move. Low-demand parts that are stocked typically remain in inventory longer than parts with higher demand, leading to relatively higher total unit costs as fixed costs such as warehousing, distribution, and labor are apportioned over a smaller total quantity of parts. Because their demands are sporadic, low, and unpredictable, low-demand items that are stocked may also generate revenues slowly over long periods of time. This can understandably lead buyers to stock fewer low-demand parts—which in turn can lead to availability problems when such parts do fail.

These problems are even more complicated for the Air Force. As the costs of aircraft have increased over time and other budget priorities have emerged, the Air Force has kept its older weapon systems flying longer than their original design lives. As systems age, more parts fail for the first time, many with old or obsolete technologies, increasing the population of low-demand parts. Among aircraft systems with an average age of at least 35 years, for example, are the C-5, the T-38, and the B-52, which have been operating since the 1950s and have been extensively modified over the years. Figure 1.1 shows the age distribution of many of the Air Force's major types of aircraft. By contrast, the average fleet age of U.S. airlines is 10.4 years (AirSafe.com, 2009).

---

also occurred if these failed parts were replaceable but not reparable. Only 4 percent of items accounting for the most MICAP incidents were insurance or nonstockage objective items for which probabilities of failure are considered to be zero or very low.

Many subsystems continue to be modernized and upgraded. New equipment and aircraft that are added to the fleet before older aircraft have been retired are giving the Air Force more parts in general and more low-demand parts in particular to manage. In addition, many subsystems on existing aircraft that are modernized and upgraded to provide improved capabilities or improved maintainability and cost can change fleet configurations, which means that the mix and number of parts on them expand. Spiral development of new weapon systems, such as unmanned aerial vehicles, also introduces more varied parts as designs evolve throughout the production period (Drew et al., 2005). Configuration differences across individual tail numbers for an aircraft type tend to increase the number of low-demand parts for such systems.

Cost and availability problems for low-demand parts pose challenges for the aggressive goals of the Air Force to improve its purchasing and supply chain management. AFMC, through the Air Force Global Logistics Command (AFGLSC) and its subordinate wings, has set a number of goals to achieve over the period September 2008 to September

Figure 1.1
Age Distribution of Types of Air Force Aircraft

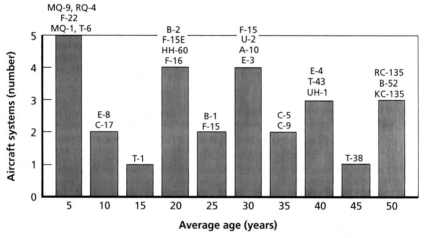

SOURCE: Logistics, Installations and Mission Support, Weapon Systems, Inventory, Average Age of Fleet, as of December 2009.
RAND MG858-1.1

2011, including a 5-percent reduction in total AFGLSC-managed inventory, improved support availability goals that increase annually, 30-percent improvement in parts availability, 50-percent reduction in contract administrative lead time, and 10-percent reduction in production lead time.[5]

Supply strategies that are shaped by proactive assessments of requirements by dimensions such as risk, strategic importance to the organization, and ease of substitution can help the management of low-demand parts. Supply strategies are based on broad, interrelated concerns of the organization, and they are guided by the strategic missions, goals, and objectives of the organization. For example, a supply strategy could be a plan for purchasing a particular group of requirements for ten to fifteen years after production.

Supply strategies consider the historical demands of a like group of requirements, their projected demands, their relative importance to the organization, and potential suppliers. They consider the end-to-end supply chain and go beyond the immediate supplier. For example, if the source of supply is a distributor, a supply strategy would also consider the relationship between the distributor and the original equipment manufacturer. It may also encompass upstream supply chains to include critical commodities or parts suppliers.

Developing proactive supply strategies for low-demand parts can help AFMC achieve its purchasing and supply chain management goals of supporting the Air Force in achieving its own goals to increase aircraft availability and reduce total costs. Developing sources of supply and repair for low-demand requirements before part demands are generated should reduce order-fulfillment time. That is, supply strategies should reduce the time needed to find, source, order, receive, and ship the part to the customer, thereby increasing the chance that when the part is needed a base or depot can provide it quickly. Working with suppliers proactively should also reduce material purchase and repair costs as long as the supplier has incentives to maintain the necessary resources to provide support and do so at a minimum total cost.

---

[5] The period of performance is 1,000 days, which began September 2008 and ends September 2011. These goals map to the AFGLSC Campaign Plan, which aligns with Air Force and AFMC goals and priorities. See AFGLSC (2010) and McCoy et al. (2010).

In this monograph, we consider possible supply strategies for Air Force management of low-demand parts. While supply chain management, supply chain performance, and inventory policies are important for these kinds of parts, the aim of this study was to look "upstream" at improvements in the ways that commercial firms and the Air Force acquire low-demand parts from their suppliers. To do this, in the next section we consider the general characteristics of low-demand parts.

Defining "Low-Demand" Service Parts

We began this research by reviewing the business, academic, and defense literatures on best practices for developing supply strategies for low-demand parts.[6] We found few studies that focused on low-demand service parts and still fewer that describe best purchasing and supply management practices for them. Most of the literature we cite comes from purchasing and supply management publications and material adapted from other, similar RAND studies (Moore et al., 2002). The literature review and commercial aftermarket services conferences helped us identify companies actively trying to improve the purchasing and supply management of their low-demand service parts.

In both our literature review and interviews with Air Force and commercial firm personnel, we found no universally accepted definition of a low-demand part. Three groups of personnel who play a role with managing such parts say they use different methods for identifying items with low demands.

Inventory and AFMC production management specialists prefer using the identifiers INS (insurance) or NSO (nonstockage objective) to characterize low-demand items. AFMC identifies INS items as those with no projected probabilities of failure but which, if failing or damaged, would require an urgent replacement to restore an aircraft to mission capability. A cockpit windshield is an example of an insurance part. AFMC identifies NSO essential items as those with very low, sporadic demands or with probabilities of failure so low they often have

---

[6] Many private firms refer to such parts as "slow moving," "non-routine," or "unscheduled." The reader should consider the term "low-demand parts" to be synonymous with such parts.

failure forecasts of zero. Rather than having zero stock because of very small demand probabilities, supply chain managers might use heuristics to set inventory requirements and stock levels of these parts.

Contracting personnel told us they use both INS or NSO characteristics and frequency characteristics to define low-demand items. Engineers, when determining if an item is low demand, reported using three characteristics, some overlapping: if an item (1) fails or is ordered very infrequently, (2) its lack of stockage could affect an aircraft's MICAP status or lead to an AWP (awaiting parts) incident, or (3) it is an INS or NSO item.

The commercial firms that we interviewed generally define low-demand items based on a low ordering frequency from their customers, who may not be the actual end-users. They do so in part because they often have visibility only on orders from distributors, not on the demand from end-users who order from these distributors. This lack of visibility on end-users or to their "installed base" precludes defining low demand based on equipment downtime or on characterizations similar to the INS or NSO designations used by the military.

Defining low-demand items by the frequency of orders for them requires setting a definition threshold, which the companies we researched set empirically. These thresholds, and the parts they include, are reviewed periodically. Thresholds can vary substantially. For example, a 1996 DLA report defined low-demand items as those with less than four requisitions or twelve total units requisitioned in the past year[7] (Hobbs, 1996), while a 2002 Air Force Logistics Management Agency report (Parrish et al., 2002) identified low-demand items, specifically NSO items, as those with no more than ten demands per year. Similarly, one representative from industry whom we interviewed told of using a threshold of one demand per year to identify low-demand items, while another cited a threshold of fewer than ten demands per month and a third cited the combination of variation and order frequency to define low demand.

---

[7] The 1996 DLA report also indicated that 80 percent of DLA's items at the time fit the low-demand definition and that about 50 percent of its inventory investment was also in low-demand items.

Definitions of low-demand parts can differ across industries or types of equipment. For example, the order threshold for defining a low-demand part for snow removal equipment used occasionally may be different and lower than that for a part in an airport radar system or medical diagnostics equipment that operates continuously.

How Can the Air Force Best Manage Low-Demand Parts?

In this monograph we analyze low-demand parts in the Air Force context and how the service might best be able to develop supply strategies for them. We do so by reviewing the general types of low-demand items the Air Force has, the best commercial practices that are used to manage such items, and the ways these practices could be applied to the Air Force.

In Chapter Two we review data on the types of items that constitute low-demand parts, the dollars that are spent on them, and the suppliers that provide them. We assess two different types of low-demand items. First, we use an ordering threshold to assess low-demand items in general and then to assess the characteristics of items whose annual orders fall below this threshold, including the spend on them. Second, we assess the characteristics of INS and NSO parts that lead to MICAP and AWP incidents. We prefer the requisition threshold for defining low-demand parts, because many INS and NSO parts are, in fact, ordered rather frequently. Nevertheless, for reasons we later discuss pertaining to data issues, using a threshold definition was not feasible for MICAP and AWP data.

In Chapter Three we discuss some of the best commercial practices that have been used in purchasing and supply chain management of low-demand parts. The research for this chapter is based on a series of interviews we conducted with industry representatives to identify best practices used by companies that were moving from a reactive to a proactive strategy for purchasing new and repairs of low-demand parts.

In Chapter Four we discuss how best practices can be applied to the purchasing and supply chain management of low-demand parts for the Air Force. We present our implications and conclusions in Chapter Five.

CHAPTER TWO

Low-Demand Service Parts for the Air Force

In this chapter we analyze low-demand Air Force orders and purchases for low-demand service parts, including how much the Air Force spends on these items and their relative importance. We show prospective target areas where the Air Force could begin applying best supply strategy practices for low-demand parts and identify companies that, as measured by dollars, sell most of the Air Force's low-demand parts.

Among our principal findings:

- most low-demand parts are in just a few federal supply groups (FSGs)
- low-demand parts represent a small number of total requisitions but a large number of parts that are requisitioned; they make up 14 percent of what the Air Force spends for parts it manages
- many low-demand parts are supplied by original equipment manufacturers (OEMs)
- most dollars for low-demand parts are spent with a small number of companies.

For our general analysis, we defined "low demand" by a requisition threshold, specifically parts for which there were six or fewer requisitions per year. By using such a threshold, we were able to link low-demand (and other) parts to spend data. Unfortunately, we were not able to adequately link the requisition data to MICAP or AWP data; many parts associated with such incidents were not found in the requisition data, which were used to identify low-demand parts. We hypothesize that some of the mismatch may stem from parts removed

Table 2.1
Synopsis of the Data Used in Our Analyses

Data	Dates	Number of Observations	Number of NIINs
Requisitions	FY 2002 to FY 2004	2,020,824	56,186
MICAP incidents	FY 1999 to mid-FY 2005	2,225,922	228,389
AWP incidents	March 2002 to March 2004	393,797	45,212
Spend	FY 2001 to FY 2003	68,238 transactions (on 23,204 contracts)	19,011

NOTE: These data contained the most current information available at the time we conducted the analyses for each data set. "Requisitions," "AWP incidents," and "spend" considered only Air Force–managed items. "MICAP incidents" were based on a larger set of NIINs managed by the Air Force, DLA, or other military services.

versus those ordered (newer versions are typically purchased) and from higher-level parts that are removed and repaired with lower-level purchased parts. For MICAP and AWP data, we used INS or NSO designation for such parts to identify low-demand parts.

For our analyses, we used data extracted from the Global Combat Support System–Air Force (GCSS-AF) for requisitions, contracts, and AWP and MICAP incidents. Table 2.1 briefly describes these data. The appendix describes in more detail how we constructed the samples for our analyses.

We first use requisitions data to identify parts with fewer than six annual requisitions. We then analyze spend data on these low-demand items (not all of which were purchased in the years for which we analyze spend data). Finally, we analyze AWP and MICAP data with respect to INS and NSO items.

Requisitions Data

Altogether, we assessed data for about two million requisitions between FY 2002 and FY 2004, identifying 40,927 out of a total of 56,186 NIINs with six or fewer annual requisitions per year. We were unable to include NIINs with zero demands over the study period.[1] Defining low demand by the number of annual requisitions is the approach most cited by representatives of commercial firms whom we interviewed and has, as noted, been used in some previous analyses of low-demand items for the Air Force.

In defining low demand, we considered the quantity of items ordered in a requisition but did not include it in our definition because the requisitions for most low-demand items were not noticeably different from items with more requisitions. We analyzed quantities per requisition for parts as a function of the frequency of their annual orders. Data on NIIN quantities in a requisition show that most (54 percent) are for an average of less than two, and that about two in three (68 percent) are for an average of less than five.

Our threshold of six requisitions was based on our observation that the rate of change in the number of orders for parts seemed to taper off after about six requisitions per year. We noted that after six requisitions per year, fewer NIINs were associated with each successive category, indicating a broad distribution of requisitions per year for other NIINs ordered more frequently.[2] We used requisitions data to assess the distribution of all items by the number of requisitions for them, the types of items that constitute low-demand items, and the leading requesters of low-demand items.

Items by Frequency of Requisition

Figure 2.1 shows the distribution of the number of all Air Force weapon-system NIINs by the annual number of requisitions for them from FY 2002 to FY 2004. Of the NIINs requisitioned during this time, 39 percent had, on average, less than one requisition per year.

---

[1] The contract and requisition data used in these analyses did not overlap well with Air Force requirements data, which prevented us from deriving a single list of NIINs managed by the Air Force that would have included the low-demand NIINs identified in the requisition data, contract NIINs, and zero-demand NIINs.

[2] One NIIN was ordered on average 4,012 times per year over the three-year analysis period.

Figure 2.1
Distribution of NIINs by Number of Annual Requisitions,
FY 2002 to FY 2004

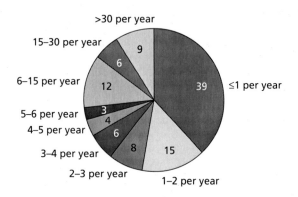

SOURCE: FY 2002 to FY 2004 Air Force requisition, Strategic Sourcing Analysis Tool.
NOTE: Numbers on pie chart represent percents. Sum of percents may not equal 100
due to rounding.
RAND *MG858-2.1*

Altogether, 73 percent of NIINs were low demand, or had, on average, six or fewer requisitions per year.[3]

The distribution of NIINs by requisitions, and of requisitions by NIINs, appears to follow a Pareto principle. Overall, 85 percent of the NIINs account for only 20 percent of requisitions (and, conversely, 80 percent of requisitions are for only 15 percent of the NIINs). More specifically, we found that low-demand NIINs comprised 10.3 percent of all requisitions between FY 2002 and FY 2004. This pattern conforms to one identified by many representatives of the companies we interviewed, who said an "80-20" rule applied to their orders as well.

As indicated, we did not include quantities ordered per requisition in our low-demand definition. Figure 2.2 shows that the quantities per requisition for low-demand NIINs were not noticeably different from

---

[3] The sum of the percent of NIINs that have six or fewer requisitions per year is actually 73, though because of rounding in the figure it appears to be 75 percent. Also, those NIINs that have less than one requisition per year may have had only one or two requisitions over the three-year period.

Figure 2.2
Average Number of Annual Requisitions of NIINs (percent),
FY 2002 to FY 2004

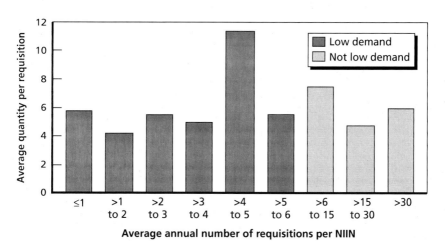

SOURCE: FY 2002 to FY 2004 Air Force requisition, Strategic Sourcing Analysis Tool.
RAND *MG858-2.2*

other NIINs that were ordered more frequently on an annual basis. Though NIINs that are ordered on average four to five times per year appear to be different, they make up only 4 percent of all NIINs and almost 5 percent of low-demand NIINs.

Types and Requesters of Low-Demand Items

Knowing the types of low-demand items that are requested can help the Air Force better target strategies for acquiring them. We analyzed data on the types of low-demand parts requisitioned and purchased as well as requestors of such items. (We later assess data by FSG in INS and NSO data on items that were used to resolve AWP and MICAP incidents.)

Table 2.2 shows the top five FSGs for requisitions of low-demand items. An FSG, defined by the first two positions of the federal supply class (FSC), permits an analysis of NIINs by category type. The FSGs in Table 2.2 comprise almost 65 percent of all Air Force

Table 2.2
Top Five FSGs for Low-Demand Requisition Items, FY 2002 to FY 2004

Rank	FSG	FSG Name	Percent
1	59	Electric and electronic equipment components	30.4
2	15	Aircraft and airframe structural components	9.8
3	66	Instruments and laboratory equipment	9.0
4	58	Communications, detection and coherent radiation equipment	8.0
5	61	Electric wire, and power and distribution equipment	7.6
		Others	35.2
		Total	100.0

sustainment low-demand items, with one, electric and electronic equipment components, comprising more than 30 percent of such requisitions.

While the components and designs among components in any similar commodity grouping will be different, any preponderance of demands for low-demand items suggests opportunities for exploring what is common across these parts to develop supply strategies. For example, are these components similar to items with high and medium demands? Are they produced by the same supplier? Could they be consolidated to single contracts? Should the Air Force directly contract for the long-lead-time raw materials that could be used by various suppliers for similar technologies?

Low-demand requisitions are made by hundreds of Air Force locations as identified by Department of Defense Activity Address Codes (DoDAACs). Table 2.3 shows the locations with the most requisitions for low-demand items as well as the locations with the most requisitions for *all* items regardless of demand.

Requisitions for low-demand items are more concentrated than those for all items. Nearly half of all requisitions for low-demand items come from the top five locations requesting such items, ranked in Table 2.3 in descending order, with the top three locations accounting for nearly 46 percent of such requisitions. By contrast, the top five locations for requisitions of all items account for only 33 percent of such requisitions.

Table 2.3
Top Five Air Force Locations with the Most Requisitions for Low-Demand
and for All Items, FY 2002 to FY 2004

Low-Demand Items		All Items	
Location	Percent of Requisitions	Location	Percent of Requisitions
Warner Robins ALC	16.0	Oklahoma City ALC	13.2
Oklahoma City ALC	15.5	Warner Robins ALC	8.2
Ogden ALC	14.2	Ogden ALC	7.0
Dyess AFB	1.9	Langley AFB	2.7
Ellsworth AFB	1.8	Eglin AFB	2.0
All others	50.7	All others	66.9
Total	100.0[a]	Total	100.0

[a] Due to rounding, percents appear to sum to more than 100.

Warner Robins ALC, which repairs and overhauls C-5, C-130, and F-15 aircraft and ground and support equipment, has the most requisitions for low-demand items, as well as the second-most overall requisitions. Oklahoma City ALC, which manages engines, tankers, and electronic surveillance aircraft, had the second-most requisitions for low-demand items. Ogden ALC, which is responsible for support of tactical aircraft, such as the F-16, and commodities, such as common avionics and wheels and brakes, had the third-most low-demand requisitions, and well as the third-most overall requisitions. Two Air Force bases, Dyess and Ellsworth, round out the top five locations with the most requisitions for low-demand items, each accounting for nearly 2 percent of such requisitions. Both bases have relatively small bomber fleets, including B-1B aircraft. Altogether, data on requisitions for low-demand items by location suggest that if the Air Force needs to improve management of these parts, it can begin by focusing on the three ALCs and then apply lessons learned there to other locations. Having many low-demand items at only three locations offers opportunities to work more closely with the suppliers of these parts.[4] It is when low demands

---

[4] Aggressive emphasis on reducing administrative and production lead times can help mitigate the impact of errors in forecasting, which are bound to arise with low-demand items because they have variable demands.

can be viewed as similar or in larger groupings that proactive supply strategies make more economic sense.

Spend Data

For spend data on low-demand items, we extracted GCSS-AF contract data for purchases the Air Force made between FY 2001 and FY 2003 for items it manages.[5] Each contract transaction recorded in these data contains the NIIN, the dollar value, the quantity of items being supplied, the unit price, and whether the transaction was sole source. We analyze as low-demand items those that were also identified as low-demand NIINs in our analysis of requisition data. Using the Consumer Price Index for "All Urban Consumers, U.S. All Items," we adjusted all dollar figures in the contact data analysis to reflect constant values for FY 2003.

Spend analyses are conducted to identify prospective targets of opportunity for reducing total costs and improving performance by increasing leverage with suppliers and by working with key suppliers to further reduce total costs and improve. They are meant to inform where to focus one's future efforts (where the greatest payoff is likely to be in developing better supplier relationships). Spend analyses identify how much is spent in certain commodity groups and with specific suppliers.

Suppliers of Low-Demand Items

Most parent companies of suppliers of low-demand items are original equipment manufacturers, such as United Technologies Corporation (UTC), Raytheon, Boeing, Honeywell, Lockheed Martin, and Northrop Grumman, which supply original parts and equipment. This suggests that improved low-demand supply strategies could begin with those manufacturers and be part of their strategic sourcing contracts.

---

[5]   During the study period the most recent data in the Strategic Sourcing Analysis Tool were for FY 2001 to FY 2003.

Table 2.4 shows the top companies as ranked by the dollar value of low-demand items they provide to the Air Force. It also lists information on total spending for spare and repair parts, the number of low-demand contracts (that is, the number of contracts with at least one low-demand item) and total contracts for spare and repair parts, the percent of spare and repair contract dollars that are for low-demand items, the percent of spare and repair contracts with a low-demand item, the percent of dollars for low-demand items that are for sole-source items, and the percent of all dollars for spare and repair items that are for sole-source goods, that the Air Force has with each firm. We list the top twelve companies because they include two additional companies that are among the top spend for both low-demand items and all items. We review below some highlights for individual firms.

The Air Force purchased $314 million in low-demand spares or component repairs from UTC between FY 2001 and FY 2003, more than it purchased from any other provider. UTC was also the single largest provider of all spare and repair items to the Air Force during this time, accounting for 18 percent of all dollars the Air Force spent on spare and repair items. Overall, 22 percent of UTC contracts had at least one low-demand item.

The second-biggest Air Force supplier of low-demand parts was Engineered Support Systems Inc., which provided $211 million in low-demand parts between FY 2001 and FY 2003.[6] These accounted for nearly 56 percent of all dollars the Air Force spent with this firm on spare and repair items. The second-biggest supplier of all spare and repair parts to the Air Force was Lockheed Martin. Only 4 percent of dollars the Air Force spent with Lockheed Martin was for low-demand parts, but this still placed Lockheed among the top ten suppliers of low-demand parts to the Air Force. In fact, the top twelve suppliers of low-demand spare and repair parts to the Air Force included the top ten suppliers of parts overall to the Air Force. This indicates that many of the firms to which the Air Force should target purchasing and

---

[6] In 2005, Engineered Support Systems Inc. (ESSI) was acquired by DRS Technologies, which is a large business. ESSI's emphasis on support was reportedly a major reason for DRS's interest in acquiring the company (Terry, 2005).

Table 2.4
Twelve Companies with the Most Low-Demand Spend, FY 2001 to FY 2003

Rank Low-Demand $s	Rank Total $s	Parent Company	$M Low Demand	$M Total for Spare and Repair Items	Number of Contracts for at Least One Low-Demand Item	Total Number of Contracts for Spare and Repair Items	Percent of Total $s That Are for Low-Demand Items	Percent of Contracts with a Low-Demand Item	Percent of Low-Demand $s Sole Source	Percent of Total $s That Are Sole Source
1	1	United Technologies	314	2,197	350	1,612	14	22	96	89
2	7	Engineered Support Systems	211	378	24	172	56	14	99	59
3	5	A1A Enterprises	130	639	704	1,346	20	52	23	39
4	3	Raytheon	79	710	115	260	11	44	24	19
5	8	Boeing	64	295	138	384	22	36	81	63
6	14	Mine Safety Appliances	51	128	7	11	40	64	100	100
7	4	Honeywell	44	661	253	902	7	28	83	75
8	2	Lockheed Martin	32	786	116	314	4	37	89	53
9	10	Northrop Grumman	29	265	208	533	11	39	60	63
10	38	ATAP	28	32	29	36	88	81	24	24
11	6	Goodrich	22	476	182	654	5	28	80	59
12	9	BAE Systems	15	272	165	414	5	40	66	58
		Others	703	5,079	5,422	16,547	14	33	49	46
		Total	1,722	11,918	7,713	23,185	14	33	64	58

NOTE: Low demand is defined as having up to and including six requisitions per year over the period FY 2002 to FY 2004.

supply chain management initiatives are also those to which it should target initiatives to improve proactive supply strategies for low-demand parts.

Two of the top twelve suppliers of low-demand parts are not among the top twelve overall suppliers of spare and repair parts. These are ATAP and Mine Safety Appliances.[7] Both firms, along with Engineered Support Systems, have a sizable portion of their sales of spare and repair parts to the Air Force in low-demand parts: 88 percent for ATAP, 56 percent for Engineered Support Systems, and 40 percent for Mine Safety Appliances. Firms that are leading suppliers of low-demand but not other spare and repair parts to the Air Force may indicate some proactive approaches to purchasing and supply chain management initiatives of low-demand parts. A1A Enterprises also had at least one low-demand item on 52 percent of its contracts for spare and repair parts with the Air Force.

Most low-demand items have a sole or single source.[8] Without competition or reasonable substitutes, the opportunities for gaining leverage over suppliers providing these goods may be even more limited than it is for other low-demand goods. Suppliers may be reluctant to institute any purchasing and supply chain management initiatives for these goods if such efforts require investment on their part and if the Air Force has no alternative source for them. Nevertheless, the Air Force could seek improvements here if information on supplier support

---

[7]   According to the Central Contractor Registration database (accessible via the Business Partner Network website), Mine Safety Appliances is a large business. Among other things, Mine Safety Appliances supplies gas masks to the Air Force. ATAP is a small business that operates depot overhaul facilities.

[8]   We determined whether a contract was sole or single source by matching the Strategic Sourcing Analysis Tool NIIN-level data with DD350 data, which contains information on the number of sources. Our analysis of these data does not differentiate between sole-source items (those for which there are no other viable and capable providers of requested goods and services) and single-source items (those for which a single source is selected among others for specific reasons). For simplicity we refer to them jointly as "sole-source." DD350 data come from DoD Form 350, Individual Contract Action Record. Any transactions of $25,000 or more were recorded at the time of this study. It has since been succeeded by Federal Procurement Data System–Next Generation, which requires reporting of transactions valued at $2,500 or more.

were shared with and used, legally, by the acquisition community when negotiating with these same suppliers for new business.

Spend on Differing Groups of Items

Table 2.2 above uses requisition data to show the FSGs in which the Air Force requisitions the most low-demand goods. Table 2.5 below uses data on expenditures to show the categories of low-demand goods on which the Air Force spends the most money. Not surprisingly, the five FSGs shown earlier rank among the top ten FSGs in which the Air Force spends money on low-demand goods, but there are others that rank higher by expenditures. For example, the Air Force spent more money on low-demand items in FSG 28, comprising engines, turbines, and components, than for any other, but this FSG was not among the top five with the most requisitioned items. Similarly, the Air Force spent more money on low-demand items in FSG 39, materials handling equipment, than any other but FSG 28, but this, too, was not among the top five with the most requisitioned items.

The FSGs shown in Table 2.5 comprise 83 percent of all spending for low-demand items and 77 percent of that for all spare and repair items. Engines, turbines, and components comprise 26 percent of all low-demand item spending and 36 percent of contract dollar spending for all items. Aircraft components and accessories similarly is among the highest-ranking FSGs for low-demand items but accounts for an even larger proportion of dollars spent on all spare and repair items. By contrast, materials handling equipment items comprise 13 percent of reported spending for low-demand items but only 2 percent of contract dollars for all spare and repair items. Overall, 96 percent of dollars spent for materials handling equipment spare and repair items were for low-demand items. All other FSGs listed in Table 2.5 similarly account for a higher proportion of spending on low-demand items than they do on all spare and repair items. Still, there was a great deal of overlap between the ranking of FSGs for low-demand parts and those for all spare and repair parts. Eight of the ten FSGs listed in Table 2.5 are also among the top ten FSGs ranked by contract dollars for all spare and repair parts, and all are among the top fifteen FSGs ranked by contract dollars for spare and repair parts.

Table 2.5
Types of Low-Demand Parts the Air Force Is Buying:
Top Ten FSGs by the Dollar Value of Low-Demand Items, FY 2001 to FY 2003

Rank for Low Demand $	Rank for Total $	FSG Number	FSG Name	$M Low Demand	$M Total	Percent of All Low Demand $	Percent of All $	Percent of Total $ That Are Low Demand
Low-Demand Items								
1	1	28	Engines, turbines, and components	442.1	4,312.40	25.7	36.2	10
2	10	39	Materials handling equipment	222.7	231.20	12.9	1.9	96
3	5	59	Electrical and electronic equipment components	163.2	775.60	9.5	6.5	21
4	4	58	Communication, detection and coherent radiation equipment	161.9	801.00	9.4	6.7	20
5	6	15	Aircraft and airframe structural components	101.7	682.40	5.9	5.7	15
6	13	14	Guided missiles	84.5	174.50	4.9	1.5	47
7	2	16	Aircraft components and accessories	70.3	1,327.00	4.1	11.1	5
8	7	66	Instruments and laboratory equipment	63.6	420.90	3.7	3.5	15
9	15	42	Fire fighting, rescue, and safety equipment; and environmental protection equipment and materials	59.1	152.50	3.4	1.3	38
10	9	61	Electric wire, and power and distribution equipment	56.2	258.30	3.3	2.2	22
			Others	296.8	2,781.80	17.2	23.4	11
			Total	1,722.1	11,917.60	100.0	100.0	14

How Low-Demand Items Are Purchased

We found that the Air Force tends to purchase low-demand items as needed, often using a purchase order in the same year that items were requisitioned, i.e., on contracts that are written after an actual demand rather than in advance of one. We hypothesize that this stems from a requirements process that might not have projected demands for these items and would not have recommended any future buy or repair requirements. Table 2.6 presents characteristics of low-demand and other (i.e., not low-demand) spare and repair items.

We found that 53.5 percent of the low-demand transactions occurred in the same year as the initial base award for the contract on which they were made, while only 40.9 percent of other items had transactions in the initial year of their contract. We also found that the Air Force is more likely to use short-term purchase orders for buying low-demand spare and repair items more often than for other parts. (Purchase orders valued of no more than $25,000 are used for lesser-dollar, one-time, or price catalog purchases.) We found that purchase orders made up 39.1 percent of all low-demand transactions (and 50.5 percent of contracts for low-demand items) but only 20.5 percent of transactions for other spare and repair items (and 37.9 percent of contracts for them). These data suggest, as noted, that the Air Force tends to purchase low-demand items as needed, rather than as part of a long-term supply strategy. This purchasing phenomenon may arise from the requirements process that often projects zero demands for low-demand parts (assuming they are represented) and to separate purchases in response to actual requisitions.

Table 2.6
Air Force Purchase Transactions of Low-Demand and Other Spare and Repair Items

	Total Transactions	Percent Same Year as Initial Award	Percent Purchase Orders
Low demand	12,131	53.5	39.1
All other	55,742	40.9	20.5

Low-Demand Parts in MICAP and AWP Incidents

Unfortunately, as noted, we were not able to adequately link requisition data to MICAP or AWP data. A portion of the NIINs in the MICAP and AWP data were not found in the requisitions data.[9] As a result, we were not able to use our preferred definition of a "low-demand" part—i.e., one with no more than six requisitions per year—to assess low-demand parts in MICAP and AWP incidents. Instead, we analyze those parts in MICAP and AWP incidents that were designated INS or NSO.

Our data on MICAP incidents span from October 1998 to April 2005. Each incident represents one occurrence of an aircraft part that was not available from supply and led to an aircraft not able to fly at least one of its missions.[10] MICAP parts can be isolated to one aircraft or more, for example, if more than one aircraft requires the same unavailable part. The data we analyzed yielded information on more than 200,000 NIINs needed to resolve more than 2 million MICAP incidents. Our data on AWP incidents span from March 2002 to May 2004. Again, each incident represents one occurrence of an aircraft that had to await parts that were not in local supply inventories.[11]

---

[9] Because sub-master NIINs are found in requisitions data, we linked sub-master NIINs to the MICAP and AWP NIINs, but this did not appreciably improve our "match" rate between the data sets.

[10] Through FY 2003, MICAP data included one record per month for each MICAP incident. For our analysis we used, for a given unique combination of document number, NIIN, and quantity, only the MICAP observation with the highest number of MICAP hours that also indicated ultimate completion of the incident, i.e., that mission capability was restored. This ensured that each MICAP incident had only one record in the data we analyzed and was counted only once. Some records we analyzed did not indicate completion of the incident. We assumed that all observations that lasted more than six months (or, more specifically, 183 days or 4,392 hours) were not properly coded as having in fact been resolved. This led us to delete approximately 1.6 percent of the approximately 3 million MICAP records in our initial data.

[11] As with MICAP data through FY 2003, a given AWP incident extending more than one month could have more than one record. For our analysis we used the record showing the highest number of AWP days for a given incident. We also deleted records showing AWP incidents exceeding six months, assuming these records were not properly coded as having in fact been resolved.

Table 2.7 indicates that about one in four NIINs needed to resolve MICAP or AWP incidents were INS or NSO items, and that about one in fourteen incidents involved an INS or NSO item. Regarding the causes of these incidents, we found that 84 percent of MICAP incidents attributable to an INS or NSO part were due to the part not being stocked at an operating location, compared to 13 percent of AWP incidents.[12] Among the MICAP and AWP NIINs, 61 percent of the MICAP INS/NSO (low-demand) NIINs and 21 percent of AWP INS/NSO (low-demand) NIINs were "not stocked" among all MICAP or AWP incidents for these NIINs. For some low-demand parts, it is more economical to stock parts centrally and to ship directly to the requisitioning unit ordering the low-demand part.

Table 2.8 shows the duration of MICAP and AWP incidents attributable to INS/NSO items and to other items.[13] MICAP and AWP incidents involving INS/NSO parts took more time to resolve than those involving other types of NIINs. The durations of MICAP incidents for INS/NSO NIINs, nearly 294 hours, were 24 percent longer than those involving other NIINs, while the durations of AWP incidents for INS/NSO NIINs, more than four days, were 5 percent longer than for other NIINs. This suggests that, on average, it took the

---

[12] About 12 percent of the remaining MICAP incidents for these kinds of parts were caused by ordered parts not yet received, whereas about 84 percent of AWP incidents have unknown causes.

The cause codes we considered to represent "not stocked" included "A—No demand or reparable generation before this request," "B—Past demand or reparable generation experience but [Air Force] base stockage policy precluded establishing level," "C—Single Manager/Inventory Management Specialist (SM/IMS) has determined the item should not be stocked at base level," and "D—Base decision not to stock the item." Cause codes for ordered parts that have not yet been received are "H—Less than full base stock – stock replenishment requisition exceeds priority group UMMIPS standards," and "J—Less than full base stock – stock replenishment requisition does not exceed priority group UMMIPS standards." Unknown causes are coded "M—undefined MICAP/AWP cause."

[13] As noted above, because some records did not apparently indicate the completion of a MICAP or AWP incident, we dropped records for all incidents that were longer than six months. To the extent that these cases did indeed exceed six months, our data, and Table 2.8, understate the length of MICAP and AWP incidents. Understatement and overstatement could pose problems for our analysis should they differ fundamentally for INS/NSO items and other items.

Table 2.7
Percent of INS and NSO NIINs and Incidents for MICAP and AWP Data
(FY 1999 to FY 2004 for MICAP and FY 2002 to FY 2004 for AWP)

	Low-Demand NIINs (Percent)	Low-Demand Incidents (Percent)
MICAP	23.8	6.8
AWP	25.6	7.7

Table 2.8
Average MICAP Hours and AWP Days by INS/NSO (Low-Demand) Status

	Not INS/NSO	INS/NSO	Difference (INS/NSO vs. Not INS/NSO)
Average MICAP hours	237.0	293.6	56.6[a]
Average AWP days	3.9	4.1	0.2[a]

[a] Indicates statistically significant differences at the 0.1 percent level.

Air Force longer to fulfill MICAP requisitions for low-demand NIINs than for other NIINs.

Summary Observations

As measured by number of requisitions, among the more typical low-demand items the Air Force requests are electric and electronic equipment components as well as aircraft and airframe structural components. Nearly half the demand for low-demand items is fairly concentrated, occurring at three air logistics centers. This suggests some initial target areas for improving supply strategies of low-demand items.

Spend data on these items reinforce the importance of these targets. All of the top five categories of low-demand items as measured by number of requisitions are also among the top ten FSGs as measured by the dollars the Air Force spends on them. The ten FSGs on which the most dollars for low-demand items are spent account for 83 percent of spend for low-demand items and 77 percent for all spare and repair items. This indicates that many of the items for which the Air Force should target general purchasing and supply chain management initiatives are likely the same ones for which strategies for low-demand items should be developed as well.

Spend data also show that many of the top parent suppliers of low-demand items are also the top suppliers of all spare and repair items. This is good news, because many of the firms for which the Air Force should seek to implement purchasing and supply chain management strategies are also those for which it should seek to implement strategies for low-demand items. A great deal of the spend for low-demand items among these top suppliers is for sole-source items, which may have implications for the kinds of supply strategies available to the Air Force for these items.

Data on the methods of purchasing low-demand items indicate that the Air Force likely purchases low-demand items as needed, rather than as part of a long-term supply strategy. This may have negative impacts on material availability to the extent that low-demand items take longer to receive than other orders.

In the next chapter we examine best commercial practices in developing supply strategies for low-demand parts. In the subsequent chapter we examine how these might be applied to the Air Force.

CHAPTER THREE

Best Practices in Developing Supply Strategies for Low-Demand Service Parts

To learn about best commercial practices in the private sector, we conducted interviews with representatives of nine companies that the business and academic literature or peer companies considered to be successful in developing supply strategies for low-demand parts. Representatives from each company we interviewed were working proactively to address low-demand parts problems so as to improve customer satisfaction, asset availability, and corporate profitability or returns on assets. The companies we interviewed were either manufacturers or supported expensive end-items purchased by customers who expected high equipment or part availability rates or were themselves customers for the end-items. We interviewed representatives from aerospace, automotive, heavy equipment, and electronics and communication firms.[1] Some firms were original equipment manufacturers, others were manufacturers of subsystems, and still others were end-users of major capital products.

The products of the companies we interviewed had life cycles that spanned 10 to 15 years (high technology and automotive), 25 years (commercial aerospace), and 40 to 50 years (heavy equipment, equipment motors). Each company had low-demand parts for the breadth of its products. Each one was in the process of developing supply strategies for most, if not all, of its low-demand parts. One company was extending support of low-demand parts beyond the timeframe required by

---

[1] As a condition of obtaining interviews, we agreed not to identify companies participating in this study.

regulation to expand its long-term product market share by improving brand loyalty and customer satisfaction. Another company was modifying its policy of very-long-life support to charging the customer all costs for any support provided past certain time limits.

Some of the companies participating in this study conducted little or no business with the Department of Defense (DoD). These participating companies supported expensive products with very high availability requirements and many low-demand parts and had experience directly relevant to this study. For those companies that did sell goods or services to DoD, we interviewed representatives of divisions selling commercial (rather than military) goods and services.

The companies we studied share several characteristics. All have global operations and multiple distribution or operating locations. Those that manufactured products had distributors with self-owned inventory and varying degrees of visibility by the company to the inventory of distributors or to end-customer demands. Many had responded to the economic downturn in 2001 by setting enterprise-level goals for their service operations and becoming more efficient, implementing best practices for purchasing and supply chain management and applying them also to low-demand parts. Those we interviewed said it took years to implement best practices, but that some investments showed positive returns in less than one year. Though all the companies that we interviewed in our study are seeking contract coverage for all their parts, none had yet achieved it. Some were making investments in process and technology improvements for all service parts, including low-demand parts, and expected to see measurable benefits in the near and long term. Some of these investments included new information systems and forecasting methodologies, supplier relationship management programs, centralized pooling of requirements across business units, etc.

Our interviewees told us that where a product is in its life cycle determines to a large degree the best practices and strategies that are available to assure supply of low-demand parts for it.

In general, they have found that the best time for developing supply strategies for low-demand parts is before production begins, when the manufacturer or buyer has the most leverage with suppliers. The incentives for providing long-term support of low-demand parts can be aligned better so that what the customer needs, the supplier is more willing to provide then, rather than later, when the balance of leverage shifts to the supplier. This alignment pertains to all parts, but especially to low-demand parts, for reasons we will discuss. We discuss the various supply strategies used for low-demand parts within the context of a product life cycle. Many of these reflect strategies that firms are implementing to improve support of all their parts. We begin by discussing private-sector goals that are leading firms to seek improvements in purchasing and supply chain management for their spare and repair parts, including their low-demand parts.

Private-Sector Goals

Private firms must satisfy both customers and their owners (or shareholders). One increasingly common means they use to do this is to give the issue of aftermarket services higher priority than it has had in the past. OEMs and retail companies consider customer satisfaction throughout the life of a product to be essential to luring customers back to buy new products.[2] Once a product is purchased, aftermarket services become the primary face of manufacturing or retail companies to the customer. Customers of expensive capital equipment demand good aftermarket support to operate their equipment efficiently and amortize their depreciation while maintaining a good resale value.[3] As technologies mature and gaps narrow among products, making them more similar, buyers are giving greater weight to products that provide the best total life-cycle costs.

As a result, aftermarket services are an increasingly important aspect of business for many companies, affecting their growth and profitability. The aftermarket parts business itself is estimated to gener-

---

[2] Once a product has been purchased, the OEM often becomes a sole-source supplier for parts and service similar to what happens in the DoD. Customer satisfaction becomes a competitive factor for future business. This works best in markets with a robust supply base and competition.

[3] A representative of one manufacturing company told us its customers try to quantify projected lost revenues due to equipment availability problems, depreciation costs, and still other measures when purchasing capital equipment.

ate $400 billion in revenues per year (Gallagher, Mitchke, and Rogers, 2005). Service contracts comprise just 3 to 4 percent of sales at leading electronics retailers such as Best Buy, but have operating profit margins above 50 percent (Berner, 2004). In fact, in 2004, service contracts contributed $600 million of Best Buy's $1.3 billion profit. Likewise, for heavy industrial manufacturing companies, aftermarket services contributed 25 percent of revenues but 40 to 50 percent of overall profits (Spellman, 2004). Similarly, the aftermarket service parts business has the highest profit margin for the automotive industry (Piszczalski, 2003). Service also represents 24 percent of revenue but 45 percent of profits in information technology (Chamberlain and Nunes, 2004). Within the aerospace industry, some firms have sought to sell new products at or near cost while anticipating profits from the sales of aftermarket services (Rossetti and Choi, 2005; Ashenbaum, 2006). Indeed, Boeing, which is manufacturing the B-787 Dreamliner, is offering life-cycle management support for the first time with its Goldcare Services product.

Given the focus of manufacturers and retailers on aftermarket services, end-users are also looking there to increase their own profits by reducing the costs of such services. Some companies, such as NetJets, which sells shares of aircraft for individual private use, have outsourced maintenance and repair workload, inventory management, and logistics and distribution operations to minimize total service costs and focus on core competencies. Buyers are leveraging their purchases of aftermarket services, including provision of low-demand items that may be needed for spares or repairs, with a few key suppliers using performance-based arrangements. These buyers are outsourcing larger chunks of service capability to those that offer comprehensive service packages. These performance-based arrangements range from guaranteed fill rates of spare parts and on-time delivery schedules of repaired parts to product availability such as power-by-the-hour service and operational or uptime rates.[4] In other cases, product users such as

---

[4]   Enslow (2004) reports that 70 percent of "manufacturing-intensive" companies and 56 percent of "distribution-intensive" companies have suppliers manage the inventory of their parts.

airlines have formed joint ventures, with different partners assuming primary responsibility for particular aspects of maintenance, repair, and support operations, such as inventory management, "heavy checks" of engines (including equipment overhauls), and airframe and avionics work.

Both end-users and manufacturers that we interviewed claimed that about 80 percent of the aftermarket service parts needed constitute about 20 percent of demand or orders for parts. Hence, they find themselves managing tens of thousands of parts that have no more than a few demands per year. Buyers find that proactively developing supply strategies for low-demand parts can positively affect equipment availability.

The benefits of improved supply strategies for low-demand parts can be quite substantial, because such parts purchased before demands occur usually remain in inventory longer and tie up resources that could have been used productively elsewhere. Low-demand parts have higher unit inventory management costs because their fixed management costs must be apportioned over fewer quantities. One way companies have considered cutting their aftermarket service costs is through reduced inventories of "slow-moving" parts that do not sell quickly (if at all). Not stocking such parts or stocking fewer of them means that responsive supply sources must be in place when demands arise. Interviewees from organizations that chose not to stock slow-moving parts told us they were trying to develop supply strategies with suppliers to better align risk and incentives to ensure that suppliers can be quickly acquired when demands arise at lower total cost. Spellman (2004) indicates that the typical service-level agreement for high-tech or retail support requires response times of 48 hours or less.[5]

---

[5] To the extent that the high-tech equipment is unique, such quick turnaround time may be challenging to achieve. However, response within this timeframe is necessary for the operation of certain Air Force equipment, such as the current generation of large unmanned aerial vehicles (UAVs). However, a set of tail numbers from a particular type of UAV may not necessarily have identical parts demands because they were acquired using a spiral development process. UAVs, like commercial aircraft, have unique ID numbers affixed to their tails that function like a serial number. Spiral development is an acquisition process that designs, develops, and produces new weapon systems on an incremental basis, adding capability with each spiral, which reduces risk, cost, and time to field and gives the warfighter enhanced capabilities sooner. However, each set of systems produced is slightly different from the pre-

Having supply strategies and suppliers in place for low-demand parts may be more cost-effective than holding many of these parts in inventory. If customers give suppliers responsibility for delivering parts to a schedule, suppliers can decide whether to hold the parts in inventory or produce them in response to actual demands. Suppliers say they can provide inventory for several companies more efficiently than a given company can provide for itself, and therefore help all reduce costs. (This advantage is more likely as the supplier's number of customers increases. If there is only one customer, other benefits must also be present for the supplier and its customer to both benefit. For example, supplier-managed inventories may reduce inventory costs if a performance-based contract incentivizes the supplier to purchase production materials for low-demand items to reduce production lead times.[6]) As customers give manufacturers and suppliers more aftermarket responsibilities, these providers seek to develop better supply strategies for low-demand parts to cut total costs while still meeting customer expectations for services.

Product Life-Cycle Phases

As noted, firms implementing best commercial practices have found that supply strategies for parts are best tailored to specific phases of a life cycle. This especially applies to low-demand parts. That is, where a product is in its life cycle will affect the supply strategy that is developed at any given time. Supply strategies developed earlier in a product life cycle will differ from those developed later.

Figure 3.1 illustrates notionally the three distinct life-cycle phases for a product: design, production, and postproduction. Different incentives are available in each phase for getting suppliers to assure long-term supplies of low-demand items. The phases of Air Force weapon system

---

vious set, leading to a greater variety of parts that are fewer in number (Johnson and Johnson, 2002).

[6]   Separate or weighted metrics that apply to low-demand items might be necessary to drive the right behavior.

"products" are much more complex than those of products in the commercial world. Nonetheless, these three phases illustrate supply strategies available over the product's life cycle. We review each of these below.

Design

During its first phase, design, a product moves from an ideal, concept, or requirement to a final set of plans describing how the product will be manufactured, used, and supported. (For purposes of this discussion, we consider research and development that may occur before design begins to be part of the design phase.) In the notional product life cycle above, this phase occurs between the start of design for the product and the start of its production (and aftermarket).

In this phase, plans become more concrete with detailed drawings and product specifications, material selection, and development of

Figure 3.1
Notional Display of Product Life Cycle, Including Design, Production, and Postproduction Phases

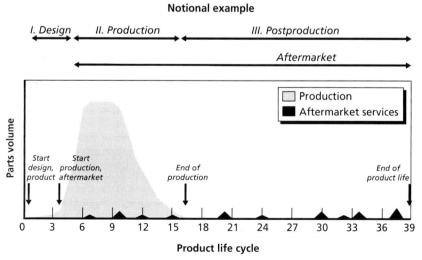

manufacturing processes. Much of the product's ultimate cost is determined in this phase. If the product uses new technologies or differs significantly from older products, a prototype may be built to test technical feasibility of design and manufacturing processes.

Estimates of aftermarket services requirements also begin in this phase. Engineers estimate failure rates through historical and engineering analyses to determine an appropriate mix and quantity for initial provisioning of spare parts. Logisticians develop support concepts for field services, levels and sources of repair, supply (inventory), and suppliers. They analyze how much repair will be done internally at organic facilities or in contracting support. They also consider the use of Contractor Logistics Support (CLS) and Performance-Based Logistics (PBL) contracts for long-term weapon system support.

Also in the design phase, the OEM determines how much product to manufacture in-house or how much to function as an integrator that assembles ready-built subsystems provided by suppliers. OEMs may develop and own the technical rights to the design of subsystems but choose not to manufacture them, instead finding suppliers to fabricate the product more efficiently. Alternatively, OEMs may find suppliers that can design and develop proprietary subsystems to meet technical and functional requirements. In this case, the supplier would own the technical rights to its subsystem and be involved in design. The extent to which suppliers invest their own capital in the design phase depends on the cost and complexity of new products.

Production

In the second phase, production, products are manufactured. The production line ramps up and operates until the manufacturer decides to end production or reallocate resources to manufacturing other products. In the notional product life cycle depicted in Figure 3.1, this phase coincides with the time period 3 to 17 (with the rate of production, and parts associated with the product, shown in the shaded area, peaking around time period 9).

Product sales, both actual and projected, influence the pace of production. Though there is uncertainty—owing to, e.g., unforeseen supply or economic shocks caused by supplier parts or capacity unavailability, labor or material shortages, or distribution bottlenecks—about the magnitude and timing of sales, once the decision on production size and schedule has been made, the manufacturer can schedule production of product parts. Lean production techniques allow suppliers to provide needed parts at a prescribed time and sequence to be assembled into final products at the same rate as production. Setting production schedules translates into certain parts requirements. Suppliers can use these schedules to plan for material and distribution capacities and material purchases.

The aftermarket services phase begins once the first products that are sold require maintenance or their parts begin to fail, typically while production is still ongoing. As the number of products in operation increases, demands for aftermarket services also increase. Service parts, including low-demand service parts, may compete for parts with new production on manufacturing lines during the production phase of a product, but they are easier to produce during this phase as well.

Postproduction
In the third phase, postproduction, production of new products has ceased and only the aftermarket phase continues. In the notional product life cycle depicted in Figure 3.1, this phase lasts from the time production ceases around time period 17 until the last products reach the end of their product life, nearly 40 time periods after the product was first designed. In this phase, business volume comprises only aftermarket services requirements. Figure 3.1 notionally shows the infrequent requirements of a particular low-demand part. In actuality, many different low-demand parts would exhibit infrequent demands over the duration of the product's operating life. Demands for low-demand parts are intermittent, variable, and economically unattractive to companies whose lines are optimized for maximum throughput.

Most companies guarantee some level of support for a period of time after production ends, but this period varies by industry and product. High-tech companies may offer support for as little as 18 months, while automotive companies generally provide aftermarket services support for 10 years after production. Some manufacturers of heavy equipment offer aftermarket services for products for 15 years following production, and then decide on a part-by-part basis which to support beyond this period. The commercial aerospace industry may offer aftermarket services for a still longer period. Boeing, for example, has publicly reported it has agreements with its avionics suppliers to provide aftermarket services as long as even only one of its products is still operating (Aeronautical Radio, Inc., 2003).

Once production has ended, supply options for low-demand parts quickly dwindle if a supply strategy is not in place. As business decreases to only aftermarket services, with no production volume, OEMs and customers start to lose their buying leverage with suppliers. Suppliers, particularly those on lower tiers who may be smaller and have fewer resources to weather lean times, may at some point exit the business altogether, especially if profitability drops below an acceptable threshold. Once the OEM or supplier decides to end support of parts that are not available elsewhere, they become obsolete. This can create problems for end-users who are still operating equipment using these parts.

The cost of providing parts with low aftermarket demands increases once production ends and they are no longer needed for manufacturing other products. If a part fails long after production ends, the OEM or supplier must locate the technical data, physical tools, equipment, and personnel required to produce the part. This may be difficult if companies have changed ownership or location. Tools also may have been discarded, and skills needed to produce the parts may have atrophied or disappeared with workforce changes. The low volume and revenue associated with low-demand parts are also unlikely to entice new suppliers to replace those leaving the field.

We next discuss the design, production, and postproduction phases in more detail, including the supply strategies available with them. As we will see, the best options are available at the earliest phases of the product life cycle. Such strategies can be proactive, addressing the issue of low demand prior to actual demands from end-users. Supply strategies in the postproduction phase are more reactive, challenging, and likely to be associated with higher costs.

Phase I, Design and Development: Minimizing Unique Parts to Reduce Total Cost and Low-Demand Parts

In general, the first, design phase of a product's life cycle can offer the most opportunities for minimizing the total number of low-demand parts. We found through interviews that companies use three broad strategies to ensure supply of low-demand parts in the design and development phase of a product's life cycle: (1) involving buyers and suppliers in the design of new systems, products, and parts, (2) reducing complexity (and hence the number of low-demand parts) by using common subsystems and parts, and (3) monitoring and managing obsolescence by identifying soon-to-be-obsolete parts, technologies, and processes. Involving buyers and suppliers in the design of new systems, products, and parts can help balance performance and cost objectives while minimizing unique parts to a specific product. Reducing complexity by using common subsystems and parts can also help minimize unique parts, and hence the ultimate number of parts that may have only low demands. Monitoring and managing obsolescence through a product's life cycle can help avoid the need to replace parts that manufacturers or suppliers might discontinue. We discuss these further below.

Strategy 1: Involve Buyers and Suppliers in the Design of New Systems, Products, and Parts

This strategy focuses first on buyers and then on suppliers. The first part of this strategy is internally focused on the internal product development processes. After discussing it, we then examine the second part of the strategy, which addresses suppliers.

Buyer Involvement. Involving an enterprise's buyers in the supply strategy process can bring to the design process important supply chain considerations that might otherwise be overlooked. The traditional approach to product design has been to use performance requirements to design a product and establish a price based on the ultimate cost of producing it. Parts were designed for performance and optimized to the particular platform. This led to a proliferation of uniquely designed parts across the portfolio of products, in turn contributing to an increase in low-demand parts.

More companies are now analyzing markets they want to target with their products and identifying performance and price ranges that suit these markets. These companies are designing products to fit performance and price targets, rather than viewing price as an output of the production process. The enterprise's buyers are critical in this process because they help ensure that a supplier's inputs conform to performance and cost targets.

Companies that bring their purchasing personnel into the design phase also do so because buyers may be familiar with past supply strategy development and support issues that could affect new programs. They also have the most recent information on prospective suppliers and market conditions. Buyers can help engineering and technical personnel incorporate supply chain quality and total cost considerations into product designs and, together with market research teams, steer designers away from parts and technologies that will soon be obsolete. If companies have organized in ways to leverage their total business, then buyers, with their enterprise-wide perspective, can also spot opportunities to improve quality or price.

Buyers can also facilitate greater use of standardized parts from fewer suppliers across systems to maximum practicable levels within OEM designs and as a condition of design contracts. Otherwise, engineering and technical buyers can help establish in design contracts with suppliers provisions for access or purchasing rights to technical data and drawings during the postproduction phase. Buyers can also help establish critical supply chain elements early. They can identify supply chain concerns or solutions to product design teams so that difficult-to-support design features can be analyzed by engineering and technical personnel.

Supplier Involvement. Some companies are also involving prospective suppliers in the design phase to improve innovation and reduce total life-cycle costs by optimizing parts availability, cost, and quality subject to design and product considerations. Suppliers know whether existing parts can fit the requirement, and thereby help reduce costs in developing new next-higher assembly components or subsystems as well as the number of low-demand parts. They can help guide engineers toward more cost-effective solutions to performance with lower expense and risk to availability and quality. Suppliers know which parts are interchangeable and can recommend ways to change designs to minimize manufacturing costs.[7] By being included in the design phase, potential suppliers of components and subsystems are able to apply their expertise to finding solutions that satisfy cost, quality, and performance concerns, sometimes by introducing new technologies to produce a part better and at less cost.

Some companies have even made their suppliers invest in new product designs, effectively making partners of them. Boeing did this with Pratt & Whitney for one of the engines that will be used on the new B-787 aircraft. The partnership spreads the risk of development and also increases incentives for suppliers to balance performance with long-term costs. Bringing suppliers into the design phase creates incentives for the same goals of looking long-term to support considerations down deeper into the supply chain to suppliers in the lower tiers. Developing a partnership can include incentives for suppliers to provide support throughout a product's life cycle, for example to recoup research and development investments they have made with OEMs.

Strategy 2: Reduce Complexity by Using Common Subsystems and Parts
The use of similar subsystems, assemblies, and parts across product lines and equipment can reduce supply chain complexity inherent to the design while extending the parts' production period, increasing total demand for them and reducing the potential number of low-demand parts. The lack of standardization of parts that could be common across products or platforms increases the total number of unique parts that must be managed, as well as supply chain complexity.[8] It can also delay, in effect, the end-of-production period of older

---

[7]   Some companies intentionally design product-unique parts to distinguish themselves competitively. By increasing the number of parts that are specific to a product, they raise the cost to entering the market for aftermarket services. Unless the product is inexpensive, the proliferation of unique parts will also raise product support costs for the customer.

[8]   Motorola reduced its product portfolio complexity in 2001 when it used more purchasing engineers in its product development and reduced its supplier base by more than two-thirds. It reduced complexity in the number of different parts and material that it needed to buy and manage (Carbone, 2001).

products, which benefit from the production of newer products using common low-demand parts.

Figure 3.2 illustrates notionally the effect of using a common part in two products with differing production runs. In this example, design of Product A begins a notional 44-period life cycle, with its production beginning in the 4th time period. Product B enters design in the 9th time period and enters production in the 12th. Product A ends production in the 16th time period, but Product B doesn't end production until the 22nd. The effect of this is to reduce the period of low demand (particularly for Product A). Aftermarket demands are depicted by a single low-demand part. Many parts that have low demands would exhibit the same infrequent low-volume demand in the postproduction phase during the product's operating life.

Companies that make incremental changes to their products to improve performance and quality may be able to use similar or even identical underlying parts of older models. The supplier can benefit from using common parts across different products. It can mitigate product risk and cost of production.

Figure 3.2
Notional Life Cycles for Products Using Common Parts

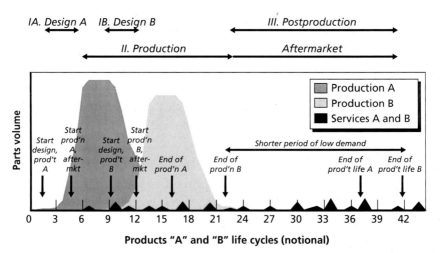

When production extends over a longer period, the supplier of parts used in several different products can benefit from a larger cumulative business volume. The broader volume reduces the average fixed cost per unit by spreading the fixed cost of designing the part and setting up the manufacturing line across more units.

The broader business volume that the use of common parts in new products permits is also likely to make the supplier consider the OEM a better customer, possibly giving it higher priority and more responsive service than buyers of smaller lots with shorter production runs. The overall aftermarket demands of these parts will increase as they are installed in more products.

Complexity can also be reduced by using technology that is already installed in other products. Another method of reducing complexity in existing systems is to upgrade subsystems simultaneously or in close succession as budgets allow. Commonly used upgraded subsystems, just like commonly used parts, reduce the number of parts unique to a product and can thereby mitigate the number of parts likely to have only low demands.

In some cases, it may be more effective to design new subsystems than to replace parts in old ones. Some OEMs wanting to be responsive to customers' support needs while reducing their own costs of providing services have found it more financially appealing to offer new subsystems to replace older ones, providing their customers business cases for how upgrades can be more cost-effective over time.[9] For example, in 2008 GE Aviation announced its intention to develop a more fuel-efficient engine on the eCore concept to replace the CFM-56-5 and -7 models on existing regional and business jets; Pratt & Whitney was developing a PurePower engine to compete in this re-engining market (Norris, 2008). Replacing subsystems may be less feasible for aerospace or industrial products with high capital costs than for electronic products. Hence, the extent to which substitution or interchangeability can

---

[9]   In the extreme, it may be more cost-effective to retire an entire system than to upgrade or maintain it. Operating old equipment while introducing new increases the number of total parts to manage, including low-demand parts.

reduce complexity and the number of low-demand parts depends on the expense of the unit being replaced.

Reducing complexity and the low-demand parts associated with it can also require the development and accurate maintenance of configuration databases that record the kinds of parts, even down to the part number or NIIN, in installed equipment. Companies are very interested in assessing their installed base of products with particular customers. Those that sell directly to end-users rather than through distributors have a better chance of end-users sharing data on operating systems and configurations. Companies with management information systems on customer relationships can assemble key data that can help them to anticipate customer demands and, in some cases, will position material in locations before demands occur. This can also be helpful for managing the distribution of low-demand parts that cannot be avoided through other strategies. Plans for positioning low-demand parts to locations that will be easier for shipping to the customer may also be developed, as could positioning them closer to express transportation and distribution systems.

To assist manufacturers or suppliers in identifying preferred parts as a means of minimizing low-demand parts, centralized parts databases can provide information on cost, reliability, preferred suppliers, preferred part numbers,[10] and obsolete part numbers. Such databases may also show the relationship that parts have with other parts, including the extent to which they can be interchanged or substituted. The value of such databases depends on the quality of their data, which means they must be updated continually with accurate information. Digitized data created from the beginning of newer systems are easier to maintain than legacy systems with multiple configurations that had their original data on paper or may be stored in incompatible electronic systems.

---

[10] As parts get older, manufacturers sometimes upgrade their materials or technologies to make them more economical, such as easier to dispose of, less expensive to manufacture, or more reliable. Parts that are more preferable to buy when purchasing a new part are referred to as "preferred part numbers." The Air Force's term for this is a "sub-group master" part.

Strategy 3: Monitor and Manage Obsolescence by Identifying Soon-to-Be-Obsolete Parts, Technologies, and Processes

Another best practice in addressing low-demand parts is to monitor and manage obsolescence by identifying parts, technologies, and processes that are being phased out or replaced. OEMs and suppliers may encourage design engineers to avoid using parts that could become difficult to obtain in the postproduction phase. Many commercial parts, especially for electronic equipment, have short shelf lives, as little as 18 months, and obsolescence that cannot be avoided is managed through upgrades or new products, e.g., cell phones. Nevertheless, some aerospace systems or industrial products may operate for decades, requiring much more attention to the management of obsolescence.

To manage obsolescence, companies create or access databases, such as those referred to in the design phase, with the most recent information on obsolete and preferred parts. This gives the company more time to decide how to adjust to discontinuation of a part. Market research and engineering teams can monitor technologies and processes for trends indicating whether groups of parts are likely to be difficult to obtain in the future. By identifying obsolescence at the earliest time possible, manufacturers can avoid or better manage future low-demand part supply problems. It may enter into discussions with the supplier about possible alternatives or look broadly at those parts that pose the greatest risk if no supply strategy is determined.

Phase II, Production and the Start of the Aftermarket: Committing Suppliers to Aftermarket Services and Ensuring Access to Technical Data

Once production, and the aftermarket, has commenced, options for ensuring availability of aftermarket parts may begin to constrict. Figure 3.1 notionally indicates this phase occurring between the 4th and 16th time periods of a product's life cycle. There are two strategies that may be developed in this phase, including committing suppliers to aftermarket services in the postproduction phase of a product's life cycle, and securing potential access to technical data. We review each of these below.

Strategy 4: Commit Suppliers to Postproduction Aftermarket Services in the Production Contract

The production phase is the best time to have aligned supplier incentives with goals for long-term support. In fact, buyers' leverage over suppliers peaks just before award of the production contract. Committing suppliers to postproduction aftermarket services in the production contract can help ensure long-term support of low-demand parts throughout the life cycle of a product. Buyers may also want to leverage the production contract award to ensure potential access to technical data for low-demand parts, which can help it develop alternative sources of supply should the original supplier exit the business or prove to be unresponsive.

Representatives of several companies we interviewed said their single most effective method of addressing low-demand parts was to secure, at the start of production, supplier commitment to aftermarket services for a specified period of time in the postproduction phase. These explicit and binding commitments were included in the production contracts. The primary incentive for suppliers to provide aftermarket support is winning the production contract as well as future production business, as companies not willing to remain in the relationship beyond production are generally not considered to be viable candidates for new programs.

The emergence of performance-based logistics also provides reasons to gain supplier commitment at the start of production to long-term support in the postproduction phase. More manufacturers are offering product support agreements at the time of purchase. This requires that the supplier network be established during production and continue after production ends.[11] Product support agreements provide incentives for manufacturers to require their suppliers to commit to aftermarket service support as part of their production contracts.

---

[11] According to Pennington (2005), in 2005 half of the total care contracts signed by GE Engine Services were agreed to as part of the contract for purchasing a new engine.

Because the availability of low-demand parts is critical to reaching performance targets, there are incentives in these agreements to include support for them.

Most of our interviewees reported that their companies had developed corporate-wide combined measures of supplier performance for production and aftermarket services. The use of such corporate-wide metrics, including performance on providing low-demand parts, provides incentives for supplier performance on these kinds of parts through links to larger, more profitable business volume. This linkage works best if performance metrics are integrated into the production award process and are taken into account when making the award. The same applies to programs that modify equipment and involve large-dollar acquisitions.

Representatives of one company we interviewed said that although their firm did not have explicit contracts for aftermarket support written during the production phase, it required suppliers to commit to providing support in the postproduction phase. It worked out the general framework of the eventual contract in the production phase and made it clear it expected suppliers to figure out how to provide parts with relatively stable (if not declining) prices during postproduction, excluding inflation. Suppliers were given notice to determine the most efficient and effective means of planning for the postproduction phase while the product was in the production phase. Later, the company wrote a contract with options to allow for negotiation and greater specificity on those aspects with uncertainty, such as performance and price targets.

In committing suppliers to postproduction aftermarket services, enterprises may encourage suppliers to have contracts flow down to lower-tier suppliers as well as seek to work with suppliers on reducing lead times and improving performance. We address each of these tactics below.

Flowing Down Long-Term Agreements with Lower-Tier Suppliers. A supplier's ability to provide low-demand parts over a long term depends on its own retention of its lower-tier suppliers. Lower-tier supply bases are typically larger than those for higher tiers, with more competitive and volatile markets as well. Our interviewees said that lower-tier companies enter and exit their markets more frequently than higher-tier suppliers do. Lower-tier suppliers are generally smaller and have fewer resources to weather fluctuations in business volume. Once lower-tier companies stop producing a low-demand part, they are not likely to ramp up again to meet increased demand. Lower-tier companies are also less concerned in satisfying the end-user, because they interact only with OEMs or other suppliers.

If OEMs and their tier-one suppliers have committed to long-term aftermarket services support, they will have incentives to secure commitments from lower-tier suppliers. Some OEMs have found that their lower-tier suppliers are often unaware of plans to provide product support for a prolonged period in the postproduction phase yet are inclined to maintain an aftermarket service relationship when approached with offers of long-term business (Berger and Williams, 2006). Communicating aftermarket plans to lower-tier companies is a necessary but not sufficient step to ensure their participation. These plans must also anticipate lower-tier supplier participation in the low-demand parts business, not just the lucrative, high-demand business.

Working with Suppliers to Reduce Lead Times and Improve Performance. OEMs have several tools to use in helping suppliers reduce lead times and improve performance. Many low-demand items are supplied by OEMs who may have manufactured the original item or are better positioned than customers to find sources of supply or to secure technical data on the item. Customers may first approach the OEM for a low-demand item because its brand is often attached to all major parts on the equipment. OEMs thus have an incentive to protect their brands by finding a source of supply satisfying their customers. OEMs are also able to leverage the business they have with suppliers to provide performance incentives for the supply of parts, including low-demand parts.

OEMs and other suppliers use objective, quantifiable performance metrics to measure whether suppliers are meeting performance or contractual goals. As a best practice, OEMs work with their suppliers to agree on the metrics and targets that OEMs expect them to achieve. These metrics, such as quality and on-time deliveries, are then measured to let both parties know how well the supplier is performing.

OEMs often collect these metrics across all activities performed by the supplier and can post results by product or aggregate them to strategic scores.

Our interviewees told us their companies use past performance metrics in conducting negotiations and awarding future contracts. The use of these metrics creates incentives for suppliers to provide support for all of their requirements, including low-demand ones. Private-sector companies do not have to ensure that these metrics are completely accurate before using them in negotiations and contract awards.[12]

Another best practice to help suppliers perform better on low-demand items is to quickly dispatch supplier development teams at the first sign of quality, cost, or performance problems. Such highlighting of problem areas can elevate the sense of urgency with the supplier and make difficult problems more likely to be resolved. OEMs are well equipped to analyze supply chain problems due to their broad experience with many suppliers, processes, and technologies. They generally have more analytical staff than their suppliers. These investigative teams include aftermarket services personnel who can specifically address low-demand support issues. Suppliers have an incentive to work with these teams because process improvements that benefit their immediate customer can also help them become more competitive elsewhere.

In some cases, investigative visits can uncover problems that OEMs inadvertently cause their suppliers and which lead to performance shortfalls. OEMs expect problems isolated to supplier practices to be remedied. Improvements in metrics will indicate whether the supplier has addressed issues that were causing supply problems.

Some OEMs offer training courses (sometimes for free) to key suppliers on applying Lean Manufacturing, Six Sigma, and other improvement practices to make their own operations more efficient and reduce variation in quality.[13] Some of our interviewees noted that

---

[12] Nonetheless, it is in the interests of suppliers to help OEMs correct inaccuracies as they occur and in the interests of the OEMs to have accurate metrics.

[13] Lean Manufacturing derives from the Toyota Production System. It is the collection of methodologies and techniques to "optimize time, human resources, assets, and productiv-

if their suppliers were not already proficient in Lean Manufacturing or Six Sigma principles, they would be required to take training as a condition for continued business.

Strategy 5: Secure Access to Technical Data in the Production Contract

Written technical data—i.e., detailed drawings, specifications, process descriptions, and technical databases—provide the blueprints needed for manufacturing new parts or repairing and servicing existing parts. Companies offering aftermarket services or customers seeking such services need to own or have access to technical data so that they can develop alternative sources of supply or repair should the original source leave the industry or its performance become unsatisfactory. Traditional firms that design products and outsource their fabrication own much of their technical data. As firms become more virtual and outsource both design and manufacturing to their suppliers, they may own less of the technical data associated with their products. Private-sector companies that incorporate technology owned by their suppliers have an analogous issue as the Air Force by not necessarily having access to these data. The best time to negotiate access to technical data is before the production contract award, when competition still exists. Once a supplier has been selected, it becomes a sole-source provider of the product's unique parts and has incentives to make access to technical data prohibitively expensive to keep third-party providers out of the aftermarket.

Technical data can be very expensive to buy, particularly if access to them has not been negotiated prior to the award of the production contract. OEMs are not inclined to share technical data without compensation, because doing so can lower the cost for competitors entering the market. As products mature and are modified, OEMs charge fees

---

ity, while improving the quality level of products and services to their customers" (Becker, undated).

Six Sigma is a methodology that uses information and statistical data analysis to identify and eliminate defects to quantities less than six standard deviations from the mean and the specification limit, e.g., fewer than 3.4 defects per million opportunities. It is used to improve any process.

to update and maintain their technical data. For very old products, first-time demands may require the OEM to look for the associated technical data that may not be digitized and available only in hard copy, if at all. A customer seeking such information during postproduction may be frustrated if the purchase of technical data was not part of the original acquisition or production contract. If the supplier's technical data are not made available to the OEM or cannot be found, the process of reengineering the part can be time-consuming and expensive.[14] This underscores the importance of acquiring access to such data no later than the production phase of an item's life cycle.

Companies can minimize these problems by stipulating in the production contract provisions for addressing the technical data or by tying future business to the transfer of current technical data. Our interviewees told us their firms had agreements for their suppliers to transfer the ownership of technical data to them after the end of production or that they otherwise expected suppliers to sell them, according to terms and conditions set in the production contract, or provide them access to technical data as a condition of good faith and doing future business together. Our interviewees added that these strategies work best when their enterprise is a key customer of the supplier. Companies sometimes buy the data and hold it in reserve to create an incentive for the supplier to perform well. Companies have the most leverage to arrange the purchase of technical data before the award of the production contract.[15]

---

[14] McDermott, Shearer, and Tomczykowski (1999, p. 1-1) note that the

Deputy Under Secretary of Defense for Logistics (DUSD(L)) indicates that the average cost to redesign a circuit card to eliminate obsolete components is $250,000 .... The Air Force is reprogramming $81 million for the F-22 program to purchase obsolete or soon-to-be-out-of-production parts and to redesign assemblies to accept commercial parts. An avionics manufacturer for the commercial airlines spent $600,000 to replace an obsolete Intel chip. The F-16 program has spent $500 million to redesign an obsolete radar. In fiscal year 1997, the KC-130F/R program spent $264,000 on a life of type (LOT) buy as a resolution for one obsolete logic device.

[15] Our interviewees said that the cost of acquiring the technical data was generally less than not having it later on. Several said they would not consider not securing access or ownership of technical data before awarding a production contract with a major supplier.

Companies told us they were subject to the same limitations as the Air Force when it comes to technical data. That is, technical data are not available in the private sector if the designs and procedures have been developed privately. Knowing the worth of these data is more straightforward in the private sector. Specified operational cost targets will constrain what suppliers can charge, and suppliers that overcharge run the risk of losing future business. Indeed, we were told that suppliers no longer wanting to manufacture very old parts may give the technical data to the OEM/customer so that the OEM can find another supplier. These private-sector practices seem rare in public-sector markets.

Phase III, Postproduction: Providing Incentives for or Developing New Sources of Parts

In the third, postproduction phase of a product's life cycle, supply options for low-demand parts rapidly dwindle if a supply strategy is not in place. Figure 3.1 indicates this phase as occurring after production ceases on an item, or about the 16th or 17th time period of its life cycle, until the end of the product's life cycle. Business volume for a spare or repair item associated with a product may now comprise only aftermarket service requirements, with few or very low demands that are highly intermittent and consequently variable. These are depicted notionally by small, black peaks in Figure 3.1. Once the OEM or suppliers decide to end support of these economically low-performing parts, they become obsolete. Lower-tier suppliers who might otherwise provide them may exit the business altogether.

Modified or updated subsystems that use common parts or fit into more than one equipment type can help extend an item's availability in the same way that overlapping production of two products, as shown in Figure 3.2, may extend production of a common item for them, or shorten the aftermarket period of low demand for an item. Business volume for modifications may be less than that of initial production because modifications may be phased in over time, as in product recalls or modernization programs, or because not all customers will modify or update product subsystems.

If an agreement is not already in place for aftermarket support, end-users should determine supplier availability for aftermarket support. More companies are trying to identify supply sources for all their products and product parts, although even those implementing best practices say they find it difficult to identify sources for many, if not every, part of their legacy systems, particularly parts with low demands. Nonetheless, the companies we interviewed were all moving in that direction. Companies may analyze potential supply problems by comparing the potential risk of not having a supply source with projected requirement quantities and total cost. In other words, firms may compare the potential problems resulting from low-demand part failures sidelining critical equipment (such as that for airport screening or medical diagnostics) with the cost of developing a supply source for some parts that may ultimately not be needed.

More companies are also analyzing how important they are to their suppliers to better understand the likely priority and responsiveness their future requirements will be given. Important customers are more likely to receive support for economically unattractive requirements, such as low-demand parts.

Should companies determine that they are relatively unimportant to a particular supplier, they may decide to move more of their future workload from that supplier to other, preferred suppliers to receive better support over a product's life cycle. This option is not always feasible during production or postproduction, or when few suppliers are available. Nonetheless, companies may find that such analysis helps them better leverage their business where they can and align supplier incentives with their own goals in seeking the best long-term support possible.

Private-sector companies identify four strategies to ensure availability of low-demand parts for aftermarket support of their installed product bases. These include providing incentives for supply of these parts, developing a new supplier of them, purchasing or retiring products for parts, or buying a lifetime supply of parts before a supplier leaves the business. We review each of them below.

Strategy 6: Provide Incentives for Supply of Low-Demand Parts

Our interviewees reported several different means their companies use to ensure that their suppliers have incentives to continue providing low-demand parts.

One such means is to guarantee future business volume across groups of low-demand parts rather than setting minimum desired quantities for each part. Companies are increasingly combining parts and other material requirements to increase their leverage over them. Low-demand parts are suitable for such a strategy to the extent that groups of them have common production equipment, skills, or material.

A second means is to combine requirements for low-demand parts with those having medium and high demands. The more important the buyer is to the supplier, the more likely that the supplier may be willing to accept an arrangement linking economically attractive business with unattractive, less-profitable business. Representatives of one company we interviewed, however, discounted this strategy, saying its suppliers would likely increase prices on all parts to minimize the risk inherent with the low-demand ones were it to pursue such a strategy.

Finally, several interviewees told us they took support of low-demand service parts into consideration when awarding new production contracts. Some buyers said they considered past performance evaluations for suppliers when awarding or negotiating new work. For one company, only a few cases of awarding business to a new supplier instead of the current one based on past performance issues sufficed for the entire supplier base to take past performance seriously. This option works best in a competitive environment.

If the buyer is not an important customer and the workload is a nuisance to the supplier or no other good alternatives currently exist, the buyer may have few options other than to pay a higher price to make the business more attractive. A proactive supply strategy would seek to avoid such a situation through the strategies outlined above or even by developing new suppliers if necessary, assuming the buyer had access to the necessary technical data and other necessary inputs.

Strategy 7: If a Supplier Does Not Currently Exist, Develop One

Some parts have been out of production so long that the supplier may no longer exist, either because the supplier has long quit production, been acquired by another company, or gone out of business altogether.

If the original supplier has been out of production for many years, it may not be interested in making the investment necessary to produce a few quantities, even if enticed with premium payments. If the part cannot be substituted with a similar one, the buyer may prefer to find or develop a new supplier for the old part rather than replace an entire subsystem or piece of equipment.

If the buyer has the necessary technical data and tooling, it may try to find a new supplier, preferably one that is already providing the enterprise other parts and has an incentive to take on additional work, even if economically unattractive. Some manufacturers have found or developed "cottage industry" suppliers that specialize in producing low-demand parts. These suppliers are able to produce parts in small lot sizes.

If the buyer does not have the necessary technical data or tooling, then its options are more limited. Companies lacking the appropriate technical data or tooling may approach the original suppliers to buy or gain access to it. Representatives of several companies told us they let suppliers know that any lack of cooperation in providing this could affect their prospects for future business.

In some cases, OEMs may purposely choose to develop a new supplier to replace an existing, more expensive source of supply. As products age and demand for their service parts begins to decrease, OEMs that may have produced parts may decide that the cost of continuing to produce parts that now have low demand is too high and will therefore try to find lower-cost suppliers specializing in low-volume, high-quality manufacturing to produce them. These low-volume suppliers may be small businesses with lower overhead costs and fewer engineering or program management personnel. Such suppliers may have demonstrated an ability to produce small, high-quality lots using specialized processes and materials, and to reduce set-up times for small lots of low-demand parts. The workforce for these suppliers must be trained to manufacture a few pieces at a time and have versatile skills. Such small businesses might also help buyers fulfill socio-economic goals, such as the federal government has, for procurement from such businesses.

Representatives of one company told us they looked for suppliers with the "right" attitude, willing to make investments, develop new capabilities, and retrain their workers. This company would work with these suppliers to bring them up to the level needed for producing low-demand parts.

To make low-demand business more attractive to new suppliers, the business can rely on economies of scope so that various low-demand parts can be produced with similar resources. To utilize this approach, requirements for low-demand parts would need to be aggregated and sourced as related groups. To be manufactured on the same line or facility, a group of parts should share similar processes, materials, and tooling, and require similar labor skills. Because of the uncertainties surrounding actual demands of particular low-demand parts, contracts with new suppliers should be negotiated for a guaranteed minimum volume for groups of parts, rather than for individual parts. In this case, price may be negotiated for a basket of parts, which as a group have a greater probability of demand than their constituent items.[16]

Buyers may also seek to increase leverage for these parts through larger, longer-term contracts. Some OEMs have been able to minimize cost to their suppliers by providing them access to their own contracts for raw materials and piece parts. When ordering off these contracts, suppliers must show that the requirement is due to the OEM's orders.

Over time, an OEM's cost per unit for managing low-demand parts may increase. This may lead it to outsource the search for and management of suppliers to enterprises that are knowledgeable about industry markets and practices for particular parts. In some cases, the

---

[16] Supply chain management of low-demand parts is complicated by their off-and-on demand. Traditional approaches of trying to forecast and buy "just enough" low-demand parts lead to excess inventory because many have no demands in any given year. Private-sector companies recognized that lean inventories—which reduce these underutilized stocks—require in-place, proactive supply strategies that can quickly react to demand or near-term forecasts. As an example, a supplier may manufacture a part up to a point and then only complete the work, such as painting it a certain color, once an actual demand is known. These kinds of parts also affect operational readiness and are important to achieving equipment performance goals.

OEM may prefer to directly contract with the new suppliers, particularly small business suppliers, to be able to claim a direct relationship with small business companies, but at the same time to outsource day-to-day management of this relationship to another company with lower overhead costs. One company said this approach permitted it to find new sources of supply for low-demand parts, reduce its management costs, free its personnel to focus on its higher-value core business, and meet a government requirement to subcontract to small business companies.

To ensure supplier diversity in outsourcing supplier management, manufacturers may write two types of contracts: one with the manager of low-demand suppliers and another with the suppliers themselves. A company may pay its manager of suppliers administrative and management fees and also pay small suppliers for the low-demand parts. Particularly for government business, this permits the company to claim subcontract status with the small business providing low-demand parts while allocating transaction costs to suppliers with lower overhead rates. Direct contracts with suppliers allow the large company to receive credit for subcontracting business to small (or diverse) suppliers. Terms and conditions in supplier contracts make them accountable to both the large company and to the supplier management firm.

Strategy 8: Purchase or Retire Whole Products Just for Parts

If parts are not available through suppliers or they are too expensive to reengineer, buyers may purchase whole products for their parts. This can be an economically viable source of low-demand or obsolete parts for old, inexpensive products or for products whose pieces of equipment have been or are expected to be retired. Buyers will incur some costs in acquiring and storing the whole products and in harvesting their parts.

Southern Air, which operates 17 B-747-200 freighter aircraft, provides an example of cannibalization for aviation parts. It buys older B-747s to cannibalize for parts, finding the practice to be more cost-effective and to result in shorter lead times than buying parts from the OEM (Dillion, 2004). The airline analyzed prospective aircraft for cannibalization beforehand, including their operating hours, age of parts, and proper Federal Aviation Administration parts certification document.

The success of this approach depends on how well the parts to be cannibalized match end-user requirements. Engine, avionics, and landing gear components are considered the most valuable parts to cannibalize. Cannibalization also entails several costs, including that for ferrying transportation and cataloging and warehousing parts. Buyers may recoup some of these costs by selling excess parts to a third party.

When only one customer owns all existing equipment, cannibalization is available only through the retirement of products. NASA had announced plans to retire, rather than to overhaul, the space shuttle Atlantis in 2008 so that it can be a "parts donor" for the remaining two shuttles in service; however, plans now call for the three orbiters to fly until 2010 (Halvorsen, 2006; Chaplain, 2008). The Air Force stores older aircraft at Davis-Monthan Air Force Base in Arizona to use as sources of reclaimed parts, drones for training, or sales to friendly foreign governments. The Aerospace Maintenance and Regeneration Center there stores about 5,000 excess aircraft, such as the F-100, F-104, F-15, F-16, C-135, UH-1, C-5, and B-52, used by the military services or other national agencies.[17]

Strategy 9: Buy Lifetime Supply of Parts Before the Supplier Exits the Business

Should a customer not wish to face the costly prospects of developing an alternative supplier or buying whole products for a few parts, it may seek to buy a lifetime supply of parts. This is typically done before a last-known supplier exits a business.

Buying a lifetime supply of low-demand parts has the direct costs of buying large amounts of material beyond any immediate need and maintaining the inventory. It has opportunity costs of spending what might be significant resources—certainly in the millions, if not billions—that might be used elsewhere. It also has costs to store the items, as DLA charges the services for warehouse storage.

---

[17] See Davis-Monthan Air Force Base (2008).

Purchasing a lifetime supply of parts, or an "end-of-life" buy, is generally considered an unattractive option due to the great uncertainties and costs involved. The choice is especially difficult for low-demand parts, given their high number of part numbers and demand uncertainties over the lifetime of a product. These estimates are especially difficult for parts that have a very-low-demand history. Some companies faced with this dilemma decide to buy an end-of-life quantity of spare parts.[18] Others decline these buys and try to manage their inventories more carefully, seeking to buy parts from competitors, or to cannibalize parts should demands exceed expectations. Companies using this strategy have to balance the cost of retaining low-demand part inventories against the chance of disappointing customers in the future.

Summary of Private-Sector Strategies

As discussed above, where a product is in its life cycle determines to a large degree the strategies available to assure supply of low-demand parts. The best options for ensuring long-term aftermarket support are in the earliest stages of a product's life cycle. Table 3.1 summarizes each strategy.

In the next chapter we discuss specific strategies the Air Force may use to manage low-demand parts in its operations.

---

[18] The Air Force has made many such buys and has many parts with zero or low demands each year. Lifetime buys are seen as necessary if the supplier is exiting the market and the Air Force does not believe it has any more cost-effective option for obtaining these parts.

Table 3.1
Low-Demand Parts Supply Strategies Availability by Product
Life-Cycle Phase

	Benefit	Challenges
Phase I: Design and Develop		
1. Involve buyers and suppliers in the design of new systems, products, and parts	Balance performance and cost objectives Minimize unique parts	Maintaining competitive sources
2. Reduce complexity by using common subsystems and parts	Minimize unique parts	Maximizing performance objectives
3. Monitor and manage obsolescence	Avoid or replace obsolete parts	Cost
Phase II: Production		
4. Commit suppliers to postproduction aftermarket services in the production contract	Ensure long-term support Life-cycle management	Contract language definition
5. Ensure potential access to technical data	OEM incentives to ensure alternative sources of supply	Cost justification in acquisition phase
Phase III: Postproduction		
6. Provide incentives for supply of low-demand parts	Maintain supply chain continuity	Cost
7. If supplier no longer exists, develop a new one	Avoid replacing equipment and subsystems	Cost Increased lead times
8. Purchase or retire whole products just for parts	Avoid replacing equipment and subsystems	Cost Increased inventory of many parts
9. Buy a lifetime supply of parts before the supplier exits the business	Avoid replacing equipment and subsystems	Cost Uncertainty

CHAPTER FOUR

Applying Best Practices for Low-Demand Parts to the Air Force

As noted in the previous chapter, best practices for low-demand parts depend greatly on where the product is in its life cycle. In this chapter we discuss how these supply strategies can be applied to the Air Force, and what strategies the Air Force is already pursuing for products in design and development, production, or postproduction.

The Air Force is different from private-sector buyers of expensive equipment. As noted earlier, many of its parts are specialized, incorporating expensive materials, are sole-source with only one buyer, have long lead times, and are subject to technology obsolescence. It has a very wide breadth of low-demand items with high unit prices and a need for reserve stocks to support a surge in operations. Having enough low-demand parts in inventory is very expensive and not economically viable as a strategy. Many low-demand parts will not be demanded, but those that are might ground aircraft if they are unavailable.

Developing proactive supply strategies for low-demand parts can occur at any time during the life cycle of a system. The companies that we interviewed were looking into developing strategies for low-demand parts for all of their systems and equipment to avoid buying inventory. Some of the solutions that apply at one phase of the life cycle are not available later on. The level of effort with which low-demand supply strategies are pursued depends on how much the low-demand part affects equipment uptime, its expense, and the cost of the supply strategies themselves. The extent to which these best practices can be applied to the Air Force environment is an essential factor.

Design and Develop

Prior to the design and development of a new or modified weapon system, the government goes through a well-structured process for setting requirements. It typically awards contracts to at least two companies for concept development, specific product research and development, and designs to meet particular requirements, before selecting one to make the product. This process may occur in stages, with companies required to include logistics and support considerations in designs that currently vary widely in their specifics.[1] The process of setting the requirement and conducting negotiations is more structured in the military than in the commercial world. Competition is always required for military programs unless particular conditions are met.

Several Air Force and DoD initiatives are changing the process for designing and developing products, including support processes for them. AFMC's Product Support Campaign initiative has been transforming the practices and processes of managing the total life cycle of weapon systems to establish and maintain their readiness and operational capability. The DoD requires that program managers consider supportability and life-cycle costs in the same way it considers performance and schedule in making program decisions. The Develop and Sustain Warfighting Systems (D&SWS) initiative is also seeking to bring life-cycle support considerations and logistics as well as ALC personnel into the acquisition phase, similar to the ways commercial firms have sought to include end-user considerations in their efforts to reduce low-demand parts (Keiper, 2006).

We discuss below more specific initiatives the Air Force and DoD have undertaken or could undertake.

---

[1] The Weapon System Acquisition Reform Act of 2009 (Public Law 111-23, signed 22 May 2009) requires the DoD to estimate the baseline for operations and sustainment costs of major defense acquisition programs, which should raise the importance of sustainment costs during acquisition. It calls for acquisition strategies that ensure competition throughout the life cycle of these major weapon systems, such as buying technical data, ensuring competition, or the option to compete, at the prime and subcontractor level, etc.

Involve Supply Chain Managers and Commodity Councils Early in Product and Parts Acquisition

ALCs could fulfill the same role in devising supply strategies for low-demand parts that buyers offer in the commercial design phase, given that ALCs are responsible for developing and managing product and logistics support of fielded systems. As a result of supporting many legacy systems, ALCs have developed expertise that could prove valuable in developing program office logistics requirements and selection criteria for suppliers of new and modified products. AFMC supply chain managers understand the effect of low-demand parts supply on weapon system availability. Members of AFMC commodity councils routinely face the challenges of assuring supplies of parts for systems in their postproduction phase. They are familiar with the performance of aftermarket suppliers and how their structure and behavior may differ from manufacturing organizations in the same firm. Their experiences can provide insights in developing supply strategies for low-demand parts early in the program for a product and in shaping the supplier aftermarket phase for the new product, system, or parts.

Reduce Complexity by Using Common Subsystems and Parts and the Proactive Management of Obsolescence

Product designs that incorporate parts used in other products can help reduce the potential number of low-demand parts. Fewer unique parts also decrease the chance that failure of any given part will affect an aircraft's mission capability. Major Air Force fighter, bomber, and space programs typically involve new technologies that go far beyond incremental changes to existing ones. As a result, new Air Force programs may have few opportunities to use standard parts or subsystems common to other programs. Still, some opportunities may exist, particularly at the subsystem and assembly levels.

The Air Force has more influence over its markets than commercial companies have over theirs, and can create requirements for new, upgraded, or standardized products that maximize common items on its platforms. In 2004, for example, the Support Equipment Commodity Council at AFMC determined that 190 unique oscilloscope stock numbers could be reduced to three common configurations (Koenig,

2005). The Air Force can encourage OEMs to avoid using parts, technologies, or processes more prone to obsolescence. On the other hand, more use of commercial-off-the-shelf (COTS) electronics would mean that obsolescence of parts would be influenced by commercial markets instead of military markets. Commercial markets are typically larger than military ones for COTS technologies and are quicker to upgrade to newer technologies over time.

DoD has developed several resources to address obsolescence directly for its legacy systems. One is the Government-Industry Data Exchange Program (GIDEP). The GIDEP is chartered by the Joint Logistics Commanders and is a clearinghouse for sharing a wide range of technical information about programs and technologies of interest to the buyers and suppliers of DoD weapon systems. Its members can access the Diminishing Manufacturing Sources Shared Data Warehouse, which is being developed by DLA and posts part numbers and national stock numbers that face obsolescence.[2]

A second resource DoD has developed to manage obsolescence of electronic parts is the Defense Microelectronics Activity (DMEA). The DMEA is a government laboratory capable of designing and developing integrated circuits and specialized microelectronic devices. It has negotiated intellectual property agreements with technology providers that allow the government to "design, prototype, and test new microelectronic components and systems" (DMEA, 2002).

The presence of sustainment considerations in the requirements process indicates that the Air Force should plan for long-term parts support as early as possible in the product life cycle. As we have noted, buyers such as the Air Force have greater leverage in ensuring the supply of low-demand parts earlier in the product life cycle. Specifying long-term logistics requirements prior to the production contract also permits suppliers to pass incentives to their lower-tier suppliers for long-term support commitment.

---

[2] Other information on the GIDEP website includes engineering data on parts, materials, and processes; metrology data for test and inspection equipment; product information data on part attribute changes; reliability and maintainability data and practices; and urgent help requests regarding sources of supply as well as other requests for information.

The emergence of CLS and PBL contracts has increased incentives for companies and their engineers to address reliability and maintainability issues with the systems supported by these contracts, including better supply strategies for the low-demand parts. OEMs are keenly interested in developing PBL arrangements for a variety of reasons, including their stable, long-term revenues. Some PBL contracts based on FAR Part 12 regulations are not subject to the same cost and pricing transparency issues as FAR Part 15 contracts and may be lucrative arrangements for OEMs, even though they may possibly reduce the Air Force's sustainment total costs for these particular aircraft systems. (Earlier we noted that many commercial manufacturers and retailers make higher profits on service arrangements than on sales of their products.) Under such arrangements, patterned after "power-by-the-hour" arrangements pioneered by Rolls-Royce, the contractor agrees to provide sufficient logistics support for a specified number of flights or missions or operating hours and the Air Force agrees to pay the contractor fixed fees for each flight (Tripp, 1998). These arrangements often include program-unique attributes, such as how replenishment spares requirements are computed and approved, who owns the spares, where they are stored, and what is considered beyond the scope of the contract (Mechem, 2006). Under such arrangements, OEMs can increase their profits if they provide parts that fail less often or parts, many of which can be low demand, to repair lines to minimize delays in completing repairs. Whether DoD is saving costs through such arrangements, however, has not yet been adequately demonstrated (Solis, 2005).

Production

The Air Force typically awards production contracts for a new product or subsystem to a single systems integrator OEM. The systems integrator OEM that is awarded the production contract uses design and manufacturing processes developed in the previous stage to begin production. Some time after the start of production, logistics capabilities develop.

In the past, the Air Force conducted most of its repairs at its own organic depots, with some contractors designated as the primary source of repair for items with proprietary technology or systems with low densities. Some contractors also functioned as secondary sources of repair for items with large variation in demands or those needed for overflow work that depots could not assume. The balance between repair work done at Air Force depots and that done by contractors has shifted over time. In 1982, DoD required the services to allocate no more than 30 percent of their repair workload to contractors; today, no more than 50 percent of all repair dollars can be spent with repair contractors, including PBL repair workloads (Cook, Ausink, and Roll, 2005).

Although a substantial amount of Air Force repair work must still be done by the Air Force itself, the Air Force also must address low-demand supply issues with its producers and suppliers for other repairs. The means available to it are similar to those available to private businesses, including developing contracts that commit the supplier to postproduction sustainment and creating incentives for the supply of low-demand parts.

Develop Contracts That Commit the Supplier to Postproduction Sustainment

As noted above, a contract that commits the contractor to postproduction support of parts, including low-demand ones, should stand alongside the production contract. OEMs have a greater incentive to commit to such support before the production contract has been awarded, particularly if they are competing with others for it. Contracts for weapon system sustainment can range from complete contractor support to support of particular subsystems, parts, or services. The key to reaching an agreement on postproduction support in the production contract is to agree to a set of rules that govern how to establish logistics support details (such as quality, price, and delivery) when more data are available on how the product is performing and its actual support requirements.

The Air Force has traditionally given higher priority to acquisition performance, cost, and schedule in determining weapon system production awards, with logistics considerations incorporated at varying degrees. This is changing.[3] The production contract for the Joint Strike Fighter was one of the first to make logistics support an important competitive criterion for the award decision. D&SWS should also elevate the importance of sustainment in acquisition processes and decisions. The best way to learn how the private sector writes these contracts is to benchmark a few companies that have been successful in providing better support of low-demand parts for less total cost.

Create Incentives for Supplying Low-Demand Parts by Tracking Supplier Performance

Performance metrics can help an enterprise evaluate suppliers' support performance as well as detect problems that need to be resolved, such as those regarding low-demand parts. Metrics that integrate support and production create incentives for OEMs and suppliers to provide support for low-demand parts.

Private-sector companies develop weights to apply to production performance variables. They do not always share these with suppliers, to prevent suppliers from "gaming" the system. Several interviewees told us that their companies base future awards on overall performance scores, such as automatic contract extensions or incentive awards. They said suppliers seem to prefer automatic contract extensions, because they reduce bidding costs and provide a strong incentive to perform well during the current phase of the contract. The Air Force has developed an extensive array of metrics to track contractor performance, although we are not aware of an integrated Air Force metric that aggregates sustainment and acquisition performance to one score.[4]

---

[3] Total Life Cycle Cost Systems Management, which requires that sustainment costs be considered more explicitly in the production phase of the acquisition of a weapon system, could make the buying of technical data more likely than in the recent past. Programs often reallocate monies earmarked for technical data to cover cost overruns during development and/or production, which essentially derails plans to buy technical data.

[4] The Joint Supplier Scorecard may offer the best opportunity for measuring suppliers across the acquisition and sustainment interface. It provides a common definition of metrics for (1) administrative lead time, (2) production lead time, (3) on-time delivery, (4) quality, (5) material readiness, and (6) cost.

Postproduction

As noted earlier, buyers have fewer options for supply strategies in the postproduction phase of a life cycle for a product than in the earlier phases. Nevertheless, some options are available, particularly for organizations such as the Air Force that have a large spend with suppliers that provide low-demand parts and could, through that spend, create incentives for suppliers to provide low-demand support more cost-effectively in the postproduction phase. We review below the options the Air Force may develop by proactively working with suppliers or by employing strategies to develop new sources of supply.

Proactively Work with Suppliers to Assure Long-Term Supply

We earlier found that most of the low-demand parts that the Air Force purchases are from its top suppliers (as ranked by dollars), and that a higher proportion of contracts for low-demand parts than for other parts are purchase orders. Using long-term contracts (e.g., 3 to 5 years) rather than purchase orders for such items can provide greater incentives for suppliers to maintain capability for supporting low-demand parts. Long-term contracts may also increase the Air Force's leverage for these items, especially if they are aggregated with more lucrative medium- and high-demand requirements. Analysis would be required to identify the right parts groupings and which organization ought to take the lead for writing the contract. The companies we interviewed said they did not see costs rise because of highly competitive markets and appropriately large market bundles. For these kinds of analyses to yield useful information, visibility to good cost and process data would also be needed.

Working with suppliers to assure supplies of low-demand parts may require the Air Force to address supply problems more proactively and to help suppliers identify root causes of quality, cost, and delivery problems. Unfortunately, contracting personnel told us, government regulations constrain the Air Force from involving itself informally in supplier operations. Contract relationships further complicate this issue, with many contracts having multiple participants (including other government agencies, such as DLA) external to the Air Force and outside its control.

Having multiple participants in supplier relationships makes it difficult for the Air Force to coordinate contractor incentives for long-term support of low-demand parts. It also makes it more difficult to develop metrics combining sustainment and acquisition measures.

Still, the Air Force can make low-demand parts and support for them more attractive to suppliers if it can link their supply to business for other, more profitable parts. In the past, some suppliers have preferred to contract with the Air Force for quantities of parts that required approval only at lower management levels, effectively increasing supplier power and profits. Commodity councils have helped reverse this trend by moving the development of supply strategies and approval of their purchases to the level of AFMC headquarters management (U.S. Air Force, 2004).

The development of commodity councils and spend analyses have made it easier for the Air Force to analyze cost and supply risk for its commodities and to identify those that are low, medium, and high demand. The establishment of commodity councils and spend analyses have also permitted the Air Force opportunities to identify where to target incentives, where feasible, for assuring supply of low-demand parts. Spend analyses coupled with low-demand parts analyses can, for example, indicate whether the Air Force should increase total business with certain suppliers that can assure supply of low-demand (and other) parts, or whether it may need to pay higher prices for such parts in order to reduce total costs for their equipment and improve overall performance.

Employ Other Strategies When Original Suppliers Are Not Available or Economical

When leverage and options are more limited, the Air Force must determine whether it has the most leverage within DoD over a supplier of low-demand parts or whether DLA or another service might. If other organizations have more leverage with the supplier, then the Air Force may wish to secure low-demand parts through these other supplier relationships by developing joint contracts that add Air Force items to larger contracts. Spotting such opportunities would require routinely analyzing DoD-wide spend at a detailed level. Now that DLA buys reparable parts for the services, the Air Force needs to be analyzing supply strategies for parts that are bought on its behalf by other organizations.

A part may, as noted earlier, be low-demand because it has failed for the first time in many years of service as a consequence of aging and long use. This may be an increasingly common problem for the Air Force as its aircraft continue to age.

If the original supplier cannot be found or is no longer producing the part, the Air Force may need to develop or qualify a new supplier. If the part is mechanical and can reasonably be reengineered, the Air Force may wish to produce it in an organic backshop. Many low-demand parts, however, such as electronic components, require a contract source of supply or repair.

If the Air Force can gain access to technical data for a low-demand part, it may be possible to develop a low-volume supplier to produce it. The Air Force will need to group these parts in ways that are economically feasible for suppliers to manufacture or repair, to be produced by companies that have demonstrated abilities to produce small lots economically.

If the parts are obsolete and technical data are not available, then the Air Force may need to cannibalize equipment for low-demand parts. Air Force policy discourages cannibalization for fleets that are not being retired or replaced.

In the past, the Air Force has sometimes purchased a lifetime supply of low-demand parts when parts become obsolete when manufacturers quit producing or supporting the part. Commercial firms, as we discussed, find this option unattractive because of the uncertainty associated with predicting lifetime needs. Some companies, as we also discussed, have opted to buy less than that needed for forecast demand, relying on careful management of remaining supplies, replacing parts with modified subsystems, or calculating the cost of maintaining additional inventories against the possibility of losing new business from disappointed current customers. Companies with PBL contracts must think about similar issues. Many legacy Air Force subsystems are too expensive to replace, making the option of replacing parts with modi-fied subsystems uneconomical in many cases.

CHAPTER FIVE

Conclusions

The Air Force's options for developing supply strategies for low-demand parts depend greatly on where its products are in their life cycles. Options dwindle as a product moves from its design phase to its production phase to its postproduction phase. The most favorable time for the Air Force to develop long-term supply strategies for low-demand parts is both during the design phase and at the beginning of the production phase before the award of the production contract. Integrating logistics plans and commitments in the production contract aligns supplier incentives with buyer goals and minimizes the total cost to the Air Force and its suppliers for assuring long-term supplies of low-demand parts. In the postproduction phase, suppliers at all levels wield more leverage than in earlier phases and are less inclined to provide less profitable low-demand parts. We summarize below actions the Air Force should take in each phase of a product's life cycle to assure support of low-demand parts.

We note that the Air Force operates many legacy aircraft, which means most of its fleet is postproduction and faces technology obsolescence, diminishing sources of supply and repair, and more low-demand failures than younger fleets. Few systems are still in the pre-award phase (e.g., the replacement tanker). Even so, many subsystems continue to be modernized and upgraded. To the extent that these programs are competed, opportunities exist to apply some of these principles even to legacy aircraft.

Phase I: Design

The Air Force could involve its sustainment representatives early in the design phase of products and parts to incorporate reliability and maintainability concerns. These representatives could also use their experience as supply chain managers and on commodity councils in supply chain considerations during the design phase. Logistics representatives could work with acquisition personnel to minimize the number of low-demand parts as part of the plan for long-term parts support as early as possible in the product cycle. Early involvement of logistics buyers can also help ensure greater consideration for supply chain issues throughout a product's life cycle.

Phase II: Production

The leverage the Air Force has over suppliers regarding supply issues for low-demand (and other) items likely peaks just before award of the production contract, when competition still exists and OEMs are more likely to agree to terms favorable to the Air Force. Including support language in the production contract can help align supplier incentives with Air Force goals and make commitment to them more likely to be honored. The time before the production contract has been awarded is also the best time for the Air Force to negotiate ownership of technical data that could prove helpful in developing alternative sources of supply (Solis, 2006).

The Air Force should require suppliers to address low-demand parts support during production (and hence the first stages of the aftermarket), when problems are easier to address than they are in postproduction when lower-tier suppliers may have quit the business and parts become more costly to support. Managing low-demand parts support during production should also address access to technical data and special tooling while both remain available with manufacturers. Private-sector companies that link support of low-demand parts to production contracts write two separate contracts: one for aftermarket support and another for production. The aftermarket support contract establishes terms and conditions and rules for supplying the parts during and after production.

Phase III: Postproduction

Many legacy systems are no longer in production and hence Air Force options for developing supply strategies for them are limited. The Air Force through its commodity councils needs to analyze comprehensively its low-demand parts to determine how to aggregate parts into more attractive packages for its suppliers. It can develop supply strategies for groups of requirements sharing similar technologies or processes, requiring certain skills, or still other similarities. Larger, related groups of requirements can increase Air Force leverage and encourage suppliers to remain in business longer.

Air Force analyses of low-demand parts can also identify what leverage it has with specific suppliers. If the Air Force has a relatively low volume of business with a supplier and cannot give it other incentives to support its parts, then it may need to pay a premium price to ensure long-term supplies. Other, less desirable options include end-of-life purchases of parts and cannibalizing whole products.

Other Considerations

In assessing supply strategies for low-demand parts, the Air Force must consider many conditions that differ from those in the private sector. For example, the Air Force has many legacy or out-of-production aircraft systems with established OEMs, limiting its ability to apply best practices for developing supply strategies for low-demand parts. Many of its low-demand parts are sole source. The market for defense systems has consolidated over the past years, resulting in fewer companies. Having fewer suppliers limits the ability of the Air Force to deny new business to suppliers that have performed poorly with low-demand parts. The development and acquisition of weapon systems in separate programs also make it difficult to achieve parts standardization and commonality. Finally, Air Force acquisition and sustainment organizations are not integrated, making it difficult to include postproduction support in the production contract, although the Integrated Life Cycle Management framework is seeking to include more consideration of logistics support of service parts in production contract language (Donley, 2009).

The source of greatest leverage in relationships the Air Force has with its suppliers may shift as a product moves through its life cycle. For some aftermarket parts that are purchased directly from tier-one or below suppliers, for example, DLA might have greater leverage with certain suppliers. The Air Force might need to find new ways to maintain relationships with its key suppliers of low-demand parts by adding low-demand parts support to its supplier relationship management discussions and summits with DLA and common key suppliers for legacy systems.[1] For new and upgraded systems, it should also be working on developing proactive strategies for securing supplies with its sustainment and acquisition communities.

Best commercial practices for low-demand service parts can support goals to increase aircraft availability rates by developing proactive supply strategies for increasing the availability of parts. They can also help AFMC meet its purchasing and supply chain management goals to reduce total sourcing cycle time, increase material supply availability, and reduce material purchase and repair costs. In sum, they can help the Air Force boost aircraft availability while simultaneously decreasing overall support costs.

---

[1]   The Air Force has greater leverage with its OEMs through its large acquisition programs, product support, and CLS and PBL contracts.

Construction of the Data Samples

Requisitions

Requisition data record Air Force unit orders for parts to be provided at the wholesale depot level. Our requisitions data span nearly three years, from FY 2002 through FY 2004. After deleting observations without a NIIN and those that were duplicate records based on unique combinations of the document number, NIIN, fiscal year, and the requisition quantity, our sample is reduced to 2,020,824 requisitions for 56,186 NIINs.[1] We use the number of requisitions for each NIIN from FY 2002 to FY 2004 as the basis for defining low-demand parts.

MICAPs

Our MICAP incidents data span nearly six years: October 1998 to April 2005. A MICAP incident is one occurrence of a MICAP. MICAPs are recorded at the NIIN level. The format of the available MICAP data changed over this period. The data through FY 2003 had one record

---

[1] The requisitions data also include information on the price of the units requisitioned. Unfortunately, the price information is poorly coded in the requisitions data, with many obvious cases in which the contract price is off by either a factor of 100 or 0.01. This was apparent from the price for the same NIIN often being 100 times or one-hundredth that of the price for that NIIN from a separate order. While we can fix some of these price discrepancies, we are probably not capturing all of the cases of an incorrect price, as some cases are ambiguous as to whether the original data are correct. Thus, the price data are not useful without first cleansing for errors, which was beyond the scope of this study.

per month for a given MICAP incident (or "document number"). A MICAP incident could have more than one record if the incident was not resolved (a part became available) within the month the MICAP incident occurred. There are likely cases where the incident was not properly coded as resolved, resulting in the creation of multiple records that appeared again in the database, because they were considered still "open" or "unresolved." Each record has an "advice code" that indicates whether the MICAP incident was starting, stopped, transferred from one end item to another, or several other choices. However, the "advice code" was missing for about one-third of the observations.

Starting with 3,014,476 MICAP records, we assumed that all observations with a MICAP that has lasted more than six months (specifically, 183 days or 4,392 hours) were not coded properly as having been resolved. While we may be deleting valid MICAP observations, we hypothesize that we are primarily deleting resolved MICAPs. These deletions did not affect the analyses described in this monograph, other than that for a comparison of MICAP hours, because we mostly considered only MICAP incidents, not duration. This "cut-off" rule of allowing no MICAP durations greater than six months led us to delete about 1.6 percent of the observations, leaving 2,966,199 MICAP records. Next, for unique combination of document number, NIIN, and quantity, we retained only the MICAP observation with the highest number of MICAP hours that had an advice code of "Z," which indicates MICAP completion. If MICAP observations for a given combination of document number, NIIN, and quantity did not have an advice code of "Z," then we took the observation with the highest number of MICAP hours because this was more likely to be the latest observation. However, some of these may be "runaway" MICAP observations that had already been resolved. These criteria left one observation per MICAP incident, or 2,255,922 MICAP incidents. These data steps resulted in over two million MICAP incidents in our final sample for 228,389 different NIINs.

AWP

We analyzed AWP incidents for the period March 2002 to May 2004. Some AWP incidents also have more than one observation. We started with 799,217 records. We excluded observations that had the calculated number of days for the AWP incident exceeding six months (specifically, 183 days). This deleted about 5.5 percent of the observations, bringing our total to 755,153 observations. We then took the observation for a given combination of document number, NIIN, and quantity that had the highest number of AWP days. This left 393,797 observations (representing one observation for every incident) for the final sample for 45,212 different NIINs.

Spend for Air Force-Managed NIINs

We extracted the contract data from the Strategic Sourcing Analyisis Tool in GCSS-AF. These data represented purchases the Air Force made for Air Force–managed NIINs between October 1, 2000, and September 30, 2003, or FY 2001 to FY 2003. We used these data to analyze low-demand parts spend.

We initially had 91,637 observations for contracts from "repairs" and 38,626 observations from contracts for "spares." Each observation represents a contract transaction, as some contracts have more than one transaction. We deleted observations with a zero or missing contract price amount or with a missing contract number, which left 71,900 repair and 33,192 spare transactions, or 105,093 total contract transactions. Deleting observations with blank contract numbers reduced the number of observations for repairs to 35,014, or 68,268 total contract transactions. This did not affect the number of spares observations.

Consolidating the 68,268 transactions into contracts produces 23,204 total contracts and 19,011 different NIINs. Of these, 65 percent had only one transaction, while 8 percent had more than 5 transactions and 32 contracts had more than 100 transactions.

Each contract transaction contains the NIIN, the dollar value, the quantity of items being supplied, the unit price, and whether that transaction was sole source. Interestingly, there are many cases where contracts had both sole-source and non-sole-source transactions.[2] We used the low-demand items identified in the requisition data analysis to define low-demand items in the contract data. We identified low-demand contract transactions.

We inflate the FY 2001 and FY 2002 contract values so that they are in FY 2003 dollars.[3] Altogether, the contract data represent $11.7 billion of item purchases, in FY 2003 dollars.

Table A.1
Top 2-Digit FSCs for Low-Demand NIINs in MICAP Data

2-Digit FSC Group	FSG Name	INS/NSO NIINs (Percent)
53	Hardware and abrasives	25.4
59	Electrical and electronic equipment components	22.7
15	Aircraft and airframe structural components	16.7
16	Aircraft components and accessories	5.4
61	Electric wire, and power and distribution equipment	4.7
	Others	25.1

Table A.2
Top 2-Digit FSCs for Low-Demand NIINs in AWP Data

2-Digit FSC Group	FSG Name	INS/NSO NIINs (Percent)
59	Electrical and electronic equipment components	33.9
53	Hardware and abrasives	17.8
16	Aircraft components and accessories	14.3
15	Aircraft and airframe structural components	7.1
29	Engine accessories	4.6
	Others	22.3

[2] A contract is determined to be "sole source" if the AMC code has the value of 3, 4, or 5.

[3] For putting the contract values in constant dollars, we use the Consumer Price Index for "All Urban Consumers, U.S. All Items" from the U.S. Department of Labor, Bureau of Labor Statistics.

Bibliography

Aeronautical Radio, Inc., *Avionics Maintenance Conference: 2003 AMC Report*, Annapolis, Md., 2003.

Air Force Global Logistics Command (AFGLSC), 448th Supply Chain Management Wing (SCMW), "1KDSP Update," briefing, Tinker Air Force Base, March 4, 2010.

Air Force Materiel Command (AFMC), Management Sciences Division, "C-5 Readiness Evaluation," Air Force Modeling and Simulation Resource Repository, Wright Patterson Air Force Base, Ohio, March 30, 2001.

AirSafe.com, "Average Fleet Age for Selected U.S. Carriers." As of November 11, 2009:
http://www.airsafe.com/events/airlines/fleetage.htm

Ashenbaum, Bryan, "Designing the Supply Chain for Production and Aftermarket Needs," CAPS Research Critical Issue Reports, February 2006.

Associated Press, "Final Space Shuttle Launch Set for May 2010," July 8, 2008.

Becker, Ronald M., "Lean Manufacturing and the Toyota Production System," SAE International, company website, undated. As of September 2009:
http://www.sae.org/manufacturing/lean/column/leanjun01.htm

Berger, Joe, and Mike Williams, "Finding Alternatives to Last-Time Buys through Effective Supplier Relationship Management/Supply Strategies," presentation to Interlog Winter, Amelia Island, Fla., January 2006.

Berner, Robert, "The Warranty Windfall," *Business Week*, December 20, 2004. As of May 14, 2007:
http://www.businessweek.com/magazine/content/04_51/b3913110_mz020.htm

Carbone, Jim, "Motorola Simplifies to Lower Cost," *Purchasing*, October 18, 2001.

Chamberlain, Joseph J., and John Nunes, "Service Parts Management: A Real Success Story," *Supply Chain Management Review*, Vol. 8, Iss. 6, September 2004, pp. 38–45.

Chaplain, Christina T., *Agency Faces Challenges Defining Scope and Costs of Space Shuttle Transition and Retirement,* GAO-08-1096, Washington, D.C.: Government Accountability Office, September 2008. As of December 2009:
http://www.gao.gov/new.items/d081096.pdf

Cook, Cynthia R., John A. Ausink, and Charles Robert Roll, Jr., *Rethinking How the Air Force Views Sustainment Surge,* Santa Monica, Calif.: RAND Corporation, MG-372-AF, 2005.
http://www.rand.org/pubs/monographs/MG372

Davis-Monthan Air Force Base, 309th Aerospace Maintenance and Regeneration Group, "309 AMARG History Factsheet," 2008, available from base website as downloadable PDF. As of September 2009:
http://www.dm.af.mil/units/amarc.asp

Defense Microelectronics Activity (DMEA), "About DMEA," 2002. As of May 22, 2007:
http://www.dmea.osd.mil/about.html

Dillion, Brian, "Highlighting Innovative Aftermarket Parts Procurement Strategies Within Southern Airlines: Disassembling Entire 747s to Ensure Parts Availability," presentation to the 2004 Interlog Aftermarket Product Support and Service Parts Logistics Conference, Huntington Beach, Calif., June 15, 2004.

Donley, Michael B., "Acquisition and Sustainment Life Cycle Management," Air Force Policy Directives 63-1 and 20-1, 3 April 2009. As of December 2009:
http://www.af.mil/shared/media/epubs/AFPD63-1.pdf

Drew, John G., Russell D. Shaver, Kristin F. Lynch, Mahyar A. Amouzegar, and Don Snyder, *Unmanned Aerial Vehicle End-to-End Support Considerations*, Santa Monica, Calif.: RAND Corporation, MG-350-AF, 2005.
http://www.rand.org/pubs/monographs/MG350/

Enslow, Beth, *Supply Chain Inventory Strategies Benchmark Report: How Inventory Misconceptions and Inertia Are Damaging Companies' Service Levels and Financial Results*, Boston, Mass.: Aberdeen Group, December 2004.

Gallagher, Tim, Mark D. Mitchke, and Matthew C. Rogers, "Profiting from Spare Parts," *McKinsey Quarterly*, February 2005.

Government-Industry Data Exchange Program (GIDEP), company website. As of September 2009:
http://www.gidep.org/

Halvorsen, Todd, "NASA Plans to Sideline Atlantis Shuttle for Parts," *USA Today*, February 20, 2006. As of May 21, 2007:
http://www.usatoday.com/tech/science/space/2006-02-20-atlantis-retirement_x.htm

Hobbs, Jeffrey J., *Analysis of Low Demand Items,* Richmond, Va.: Department of Defense Logistics Agency Operations Research Office, DLA-96-P40251, June 1996. As of September 2009:

http://www.dtic.mil/cgi-bin/GetTRDoc?AD=ADA310419&Location=U2&doc=
GetTRDoc.pdf

iSixSigma.com, company website. As of September 2009:
http://www.isixsigma.com/

Johnson, Col Wayne M., and Carl O. Johnson, "The Promise and Perils of Spiral
Acquisition: A Practical Approach to Evolutionary Acquisition," *Acquisition Review
Quarterly,* Summer 2002. As of October 9, 2009:
http://www.dau.mil/pubscats/PubsCats/AR%20Journal/arq2002/JohnsonSM2.pdf

Keiper, Matt, "The Acquisition Transformation Action Council Approves
Initiation of the Future Acquisition Team Focus Areas," *Agile Acquisition,*
February 2006. As of May 22, 2007:
https://www.safaq.hq.af.mil/news/february2006/acpo.html (link no longer available)

Koenig, Edward, "Supply Chain Council Awards for Excellence in Supply Chain
Operational Excellence," Air Force Materiel Command, Directorate of Logistics
and Sustainment, Wright Patterson Air Force Base, Ohio, February 15, 2005. As
of May 22, 2007:
http://www.acq.osd.mil/log/sci/awards/2004_award/AFMC_Directorate_for_
Logistics_and_Sustainment_Purchasing_and_Suppl_%20Chain_Management_
Commodity_Councils_Initiative.pdf

Leatham, Mark, "eLog21—Purchasing and Supply Chain Management," *Air Force
Journal of Logistics,* Vol. 27, No. 4, Winter 2003, pp. 44–46. As of May 4, 2007:
http://www.aflma.hq.af.mil/lgj/Vol_27_No4_WWW1.pdf

McCoy, Maj Gen Gary T., Lorna B. Estep, M. Scott Reynolds, Jeffrey R. Sick,
Roger D. Thrasher, and Richard D. Reed, "2010 AFGLSC Campaign Plan,"
AFGLSC, Scott Air Force Base, January 2010.

McDermott, Jack, Jennifer Shearer, and Walter Tomczykowski, "Resolution
Cost Factors for Diminishing Manufacturing Sources and Material Shortages,"
Annapolis, Md.: ARINC, May 1999. As of May 18, 2007:
http://www.gidep.org/data/dmsms/library/dmea3.pdf

Mechem, Michael, "Suppliers Sign on for 787 Goldcare," *Aviation Week & Space
Technology,* Vol. 165, No. 5, July 31, 2006, pp. 60–61.

Moore, Nancy Y., Laura H. Baldwin, Frank Camm, and Cynthia R. Cook,
*Implementing Best Purchasing and Supply Management Practices: Lessons from
Innovative Commercial Firms,* Santa Monica, Calif.: RAND Corporation,
DB-334-AF, 2002.
http://www.rand.org/pubs/documented_briefings/DB334/

Morgan, James P., "Inside Intel," *Purchasing's Book of Winners,* 1995.

Norris, Guy, "Future eCore Foundation Plan Revealed," *Aviation Week & Space
Technology,* July 13, 2008. As of 27 April 2009:

http://www.aviationweek.com/aw/generic/story_generic.
jsp?channel=awst&id=news/aw071408p2.xml

Parrish, Tony, Scott Naylor, Rachel Oates, Robert A. Nicholson, Doug Blazer, and Bob McCormick, *Air Force Regional Stockage Policy Opportunities: Part I Regional Stock Levels for Non-Communication Electronic, Low Demand Items*, AFLMA Report LS200126302, Gunter Annex, Ala.: Maxwell Air Force Base, October 2002.

Pennington, Victoria, "60 Minutes of Power," *Airfinance Journal,* Supplement, June 2005, pp. 16–18.

Piszczalski, Martin, "IT for the Aftermarket," *Automotive Design and Production*, Vol. 115, Iss. 12, December 2003, pp. 16–17.

Rossetti, Christian, and Thomas Y. Choi, "On the Dark Side of Strategic Sourcing: Experiences from the Aerospace Industry," *Academy of Management Executive*, Vol. 19, No. 1, February 2005, pp. 46–60.

Siekman, Philip, "GE Bets Big on Jet Engines; Despite Airlines' Woes, It Is Counting on Efficiently Produced New Models to Generate Decades of Profits," *Fortune Magazine*, December 30, 2002. As of November 2009: http://money.cnn.com/magazines/fortune/fortune_archive/2002/12/30/334574/index.htm

Solis, William M., *Defense Management: DoD Needs to Demonstrate That Performance-Based Logistics Contracts Are Achieving Expected Benefits*, GAO-05-966, Washington, D.C.: Government Accountability Office, September 2005. As of May 22, 2007: http://www.gao.gov/new.items/d05966.pdf

Solis, William M., *Weapons Acquisition: DoD Should Strengthen Policies for Assessing Technical Needs to Support Weapon Systems*, GAO-06-839, Washington, D.C.: Government Accountability Office, July 2006. As of May 23, 2007: http://www.gao.gov/new.items/d06839.pdf

Spellman, Dan (2004), "Strategic Inventory Management," Caterpillar Logistics FT Services, L.L.C., briefing, Philadelphia, Penn.: Council of Logistics Management Annual Conference, October 3–6, 2004.

Streeter, Melanie, "Air Force Logistics Moves into New Century with 'eLog21,'" *Air Force Print News,* December 4, 2003. As of May 4, 2007: http://findarticles.com/p/articles/mi_prfr/is_200312/ai_3726196882

Terry, John, "The Deal: Engineered Support Bought by DRS Technologies," *St. Louis Business Journal,* December 16, 2005. As of May 4, 2007: http://stlouis.bizjournals.com/stlouis/stories/2005/12/19/focus2.html

Tripp, Edward G., "Beyond Warranty," *Business and Commercial Aviation*, Vol. 83, No. 2, August 1998, pp. 88–96.

U.S. Air Force, "Commodity Council Implementation and Operations," Air Force Federal Acquisition Regulation Supplement (IAFFARS) Informational Guidance (IG) 5307.104-93, December 2, 2004.

U.S. Department of Labor, Bureau of Labor Statistics, website. As of September 2009: http://data.bls.gov/cgi-bin/surveymost?cu

Warner Robins Air Logistics Center, "Fiscal Year 00 Annual History, Chapter IV, Aircraft Management, Tiger Team," Robins Air Force Base, Georgia, 2000.